P9-DMC-357

NUCLEAR PARADOX

NUCLEAR PARADOX

Security Risks of the Peaceful Atom

Michael A. Guhin

American Enterprise Institute for Public Policy Research
Washington, D. C.

Michael A. Guhin is a private consultant on foreign policy, including international atomic energy affairs, and a former member of the National Security Council staff.

ISBN 0-8447-3204-4

Foreign Affairs Study 32, March 1976

Library of Congress Catalog Card No. 76-6139

Printed in the United States of America

CONTENTS

ACKNOWLEDGMENTS

This essay, written in association with the American Enterprise Institute for Public Policy Research, would not have been possible without the confluence of two other personally happy events. An International Affairs Fellowship from the Council on Foreign Relations and a Research Fellowship from the Washington Center of Foreign Policy Research at the Johns Hopkins School of Advanced International Studies, combined with AEI's support for this project, gave me a year of reflection and study away from the demanding pace of government service.

I am especially grateful to these three institutions for that opportunity, and to Alton Frye, Robert E. Osgood, Robert W. Tucker, David P. Calleo, and Robert J. Pranger for their kind assistance in my sabbatical year.

Many friends and colleagues provided penetrating comments and constructive criticisms of earlier drafts. For all their help I particularly want to thank David Elliott, Leon Fuerth, Victor Gilinsky, Robert Gallucci, Benjamin Huberman, Jerome Kahan, Jan Kalicki, Christopher Makins, Henry Nau, Robert J. Pranger, Charles Van Doren, and Mason Willrich.

The responsibility for the final product, however, is mine. The views expressed in the following pages are my own and do not represent the official views of the National Security Council, its staff, or any other government agency.

MICHAEL A. GUHIN

INTRODUCTION

The development and dissemination of nuclear energy for peaceful purposes present an inescapable dilemma: How does one guard against the dangers of proliferating nuclear weapon capabilities while advancing the benefits of the atom when the basic technology for both is largely the same? The United States has gone through various phases in attempting to resolve this dilemma—sometimes stressing control and at other times stressing promotion of nuclear energy. The country has, in the mid-1970s, shifted again toward increasing emphasis on nonproliferation and protective measures. As chief developer and promoter of nuclear energy, the United States shoulders much of the responsibility for protecting against the security risks inherent in the peaceful atom.

Washington has sponsored and helped achieve, with the support and contributions of several countries, an evolving political-legal structure of constraints and obligations which today includes widespread acceptance of international safeguards and the Nuclear Nonproliferation Treaty (NPT). These, along with other established controls, comprise very important policy components contributing to the nuclear nonproliferation objective of avoiding—or at least minimizing—the spread of nuclear explosive capabilities whether labeled for military or for peaceful purposes.

Significant problems and serious doubts remain, however, regarding the adequacy of existing controls in light of the nature of potential nuclear threats, national and subnational, and the rapid worldwide expansion of nuclear energy programs. The threats, detailed in later sections, boil down to the acquisition of nuclear weapon capabilities by countries not now possessing them, including countries in regions of instability and conflict, and the possession of nuclear explosives or highly radiotoxic plutonium by dissident, radical, or terrorist groups.

The emerging nuclear energy picture drives these risks home more forcefully than in the past. For fueling growing numbers of nuclear power reactors, the outlook over the next two decades encompasses the expansion of capabilities to enrich uranium (by isotope separation) and to separate plutonium from spent reactor fuel (by chemical reprocessing). Many hundreds of thousands of kilograms of weapon-usable nuclear materials, particularly plutonium, are expected to be in use, transport, or storage in various parts of the globe, with transport of significant quantities of nuclear materials running into several thousands of shipments annually.

These technological capabilities and nuclear materials also represent key preconditions for nuclear weapon development. Reprocessing or enrichment technologies, plutonium or high-enriched uranium, and the technical infrastructure in peaceful programs will not impel any country to acquire nuclear explosives. But they can, by lowering the technical and economic barriers to nuclear explosive programs, make such development easier for a country inclined in that direction, whether now or in the future. Growing amounts of highly radiotoxic and weapon-usable materials in peaceful programs can increase the opportunities for diversion or theft by a group or individual bent on nuclear terrorism or terrorism generally.

In two short years such events as the energy crisis, India's nuclear blast, Brazil's arrangements with West Germany to acquire enrichment and reprocessing capabilities, and increasing interests in nuclear energy in the Middle East have focused attention on and greatly heightened concern about the problems and prospects of controlling the security risks of the peaceful atom. Recent developments have also altered part of the nonproliferation puzzle. They have not, however, changed the fundamental problems associated with the coming of age of the "atoms for peace" programs. Whether accelerated by an energy crisis sooner, later, or not at all, the forecast growth of the nuclear energy industry would compound the intrinsic dilemma of the peaceful atom.

The United States and other countries are at a crossroads in determining whether this growth will be more or less controlled, or more or less dangerous and destabilizing. Several difficulties are at issue in pursuing a path toward significant programs and controls to reduce the proliferation and other security risks of the peaceful atom. Regardless of efforts to reduce the risks, the years ahead may well mark the addition of new members to the so-called nuclear club. The potential of subnational nuclear terrorism may become a reality. At the same time, major political, military, economic, and energy issues will continue to make heavy demands on United States foreign policy. A legitimate question at the outset, therefore, is how much priority the United States should give to antiproliferation efforts.

The dire and doomsday forecast of rapidly rising risks leading inevitably to global nuclear holocaust need not be invoked to accept the proposition that the problems of the proliferation of nuclear weapons and nuclear explosive capabilities require serious attention. Such predictions appeared when the United States ushered in the nuclear age in 1945, when the Soviet Union's first nuclear blast shook the noncommunist world in 1949, and when France and the People's Republic of China joined the nuclear ranks in 1960 and 1964 respectively. The consequences of any and all proliferation have, it seems, frequently been overdrawn in the past. This has resulted in the corrective suggestion that, depending upon the case and surrounding circumstances, a world of additional nuclear powers need not be automatically, drastically, and irreversibly destabilized.[1]

Notwithstanding any past exaggerations of inevitable consequences and the contrasting analyses, it remains extremely difficult to understand how the existence of what Secretary of State Rusk termed "additional centers of independent decision-making on the use of nuclear weapons"[2] would increase international and regional stability or contribute to peaceful relations among nations. It remains extraordinarily easy to understand how the opposite would hold true.

The emergence of several small nuclear powers may present no direct threat to the nuclear superpowers. But it would surely introduce (1) complications, as in achieving international or regional restraints on nuclear weapons, (2) some unknowns, such as the possible transfer of nuclear capability to other countries or groups, and (3) substantial risks, including an increase in the possibilities of technical accident, unauthorized use, preemptive-surgical military actions by others, and third-country provocative attack.

---

[1] See James E. Dougherty, *How to Think about Arms Control and Disarmament* (New York: Crane, Russak, and Co. for the National Strategy Information Center, 1973), pp. 145-149; Fred C. Iklé, "Nth Countries and Disarmament," *Bulletin of Atomic Scientists*, vol. 16 (December 1960), pp. 391-394; and Morton A. Kaplan, *System and Process in International Politics* (New York: John Wiley and Sons, 1957), p. 50 (both as mentioned and summarized in Dougherty); and Zbigniew Brzezinski, *The Fragile Blossom: Crisis and Change in Japan* (New York: Harper Torchbooks, 1972), pp. 136-137.

[2] Testimony of Dean Rusk, U.S. Senate, Committee on Foreign Relations, *Hearings on the Nonproliferation Treaty*, 90th Cong., 2d sess. (1968), p. 4.

Nuclear proliferation would essentially and seriously compound already existing problems:

> It would be infinitely more difficult, if not impossible, to maintain stability among a large number of nuclear powers. Local wars would take on a new dimension. Nuclear weapons would be introduced into regions where political conflict remains intense and the parties consider their vital interests overwhelmingly involved. There would, as well, be a vastly heightened risk of direct involvement of the major nuclear powers.[3]

Although the degree of danger would vary, depending upon the country or countries that went nuclear, regional risks and their global implications are potentially grave in several instances.

The situation in the Middle East is an obvious and oft-cited example. The nuclear threat in this volatile region can be summarized in the hypothesis that "if war is not curbed in the Middle East, it will eventually become nuclear."[4] The chances of its becoming so, under conflict conditions, would not likely be lessened by local nuclear weapon capabilities or by capabilities to "go nuclear" in a relatively short period of time. While the Middle East stands out today, known dangers in other regions of the world could be compounded significantly by nuclear proliferation, as could the yet-to-be-known dangers of an uncertain future.

The possibility that the complications and dangers may prove manageable in a coping world does not erase the value of avoiding them when and where this can be done. The United States and, by their own declaration, most other countries have no present interest in facilitating the spread of nuclear weapons. This alone requires some high priority to antiproliferation efforts. Such efforts need not detract attention and energy from the equally and, sometimes, much more urgent, consequential, and far-reaching issues of the times. Nor do they obviate the need for understanding how the adverse consequences of further proliferation might be minimized. But with the cooperation of others, antiproliferation efforts can influence to what degree and at what pace, if not whether, the world will have to deal with these consequences.

---

[3] Henry A. Kissinger, "An Age of Interdependence: Common Disaster or Community," address before the United Nations General Assembly (23 September 1974), in *Department of State Bulletin*, vol. 71, no. 1842 (14 October 1974), p. 501.

[4] Robert J. Pranger and Dale R. Tahtinen, *Nuclear Threat in the Middle East* (Washington, D. C.: American Enterprise Institute for Public Policy Research, July 1975), in Preface.

The purpose of this essay is to examine recent U.S. proposals [5] and other policy courses which could be pursued on an international level to reduce the national and subnational risks of the peaceful atom, along with a few basic principles for approaching these problems today and for some years down the road. The discourse concentrates on those security angles and dangers directly associated with the diffusion of nuclear materials, equipment, and technology through nuclear energy programs, while recognizing that nonproliferation of nuclear explosives has many other facets and constitutes only one piece of a larger political-security puzzle.

[5] Some of the U.S. proposals of late 1974 have, as announced just recently, resulted in agreement among the major nuclear supplier countries on "principles governing nuclear exports." See statements of the director of the Arms Control and Disarmament Agency and the director of the Department of State's Bureau of Politico-Military Affairs before the Senate Subcommittee on Arms Control, International Organizations, and Security Agreements, Committee on Foreign Relations, ACDA Press Release of 23 February 1976 and Department of State Press Release of 24 February 1976 respectively; and *New York Times*, 24 February 1976, p. 1.

The principles of the agreement, as reported, are outlined in Chapter 3. The discussion there of similar nuclear supplier controls needs to be read in light both of this significant step forward and of what more needs to be done.

1
PREVIOUS ROUNDS WITH THE NUCLEAR DILEMMA

Ever since the birth of the Manhattan Project and the destructive debut of nuclear weapons over the Japanese cities of Hiroshima and Nagasaki in August 1945, the double-edged nature of nuclear power has, to varying degrees, been understood and respected. On the military side, the question was how to avoid or reduce the dangers of the proliferation of nuclear arsenals. The United States and its two close allies in the development of the atomic bomb, Great Britain and Canada, could not monopolize the basic principles of nuclear science and related technological advances.

On the nonmilitary side, the question concerned how, if at all, potential peaceful benefits of nuclear power (particularly in generating electricity) could be attained while keeping the security risks within reasonable limits. As noted in the tripartite declaration of November 1945, the "methods and processes . . . required for industrial uses" would be largely the same as those for "the military exploitation of atomic energy." The three allies thus agreed that the establishment of effective international safeguards against military applications of nuclear power should precede any diffusion of nuclear technology and materials for potential nonmilitary or industrial purposes. Hence came their proposal for a commission on atomic energy, within the framework of the United Nations, to examine and make recommendations for controls on peaceful programs, for nuclear disarmament, and for reliable safeguards (including inspections).[1]

---

[1] See "Joint Declaration by the Heads of Government of the United States, the United Kingdom, and Canada (15 November 1945)," in *Documents on Disarmament 1945-1959*, vol. 1 (Washington, D. C.: Department of State, 1960), pp. 1-3. For the introductory sections of this chapter, the author has drawn upon and is indebted to Mason Willrich and Theodore B. Taylor, *Nuclear Theft: Risks and Safeguards*, A Report to the Ford Foundation's Energy Policy Project (Cambridge, Mass.: Ballinger, 1974), especially the "Historical Background" summary, pp. 175-192.

The Baruch Plan

The first major attempt by the United States to find answers to both questions was presented at the opening session of the United Nations Atomic Energy Commission, in June 1946, in the form of the Baruch Plan. Based upon a report directed by then Deputy Secretary of State Dean Acheson and special consultant David Lilienthal, the plan envisioned: (1) the establishment of an International Atomic Development Authority which would control, own, and manage those nuclear activities of most concern from the standpoint of their applicability to military weapon programs and would have the authority to license and inspect all other nuclear activities, and (2) the subsequent destruction of existing nuclear weapons.[2]

Nuclear disarmament and the development and dissemination of atomic energy for nonmilitary purposes were locked together in the United States proposal and conditioned on the prior establishment of effective controls and safeguards against military applications. The Soviet Union's counterproposal, presented by then Deputy Foreign Minister Andrei Gromyko, took another tack. It called for (1) the conclusion of an international convention to prohibit the production, stockpiling, and employment of atomic weapons, and (2) measures, such as further studies and recommendations, aimed toward establishing methods of observance and "a system of control."[3]

By placing a commitment to nuclear disarmament ahead of any international controls and verification machinery, the Gromyko Plan amounted to an approach already rejected by the United States—banning the bomb without "effective guarantees of security and armament limitation."[4] Two years of negotiations narrowed a few of the many differences between the approaches advanced by Washington and Moscow, but the wide gulf between the two capitals was not to be bridged.[5]

[2] See "The Baruch Plan," statement by the U.S. Representative to the United Nations Atomic Energy Commission (14 June 1946), in *Documents on Disarmament 1945-1959*, vol. 1, pp. 7-11; and "A Report on the International Control of Atomic Energy," a summary of the Acheson-Lilienthal proposal, in *Department of State Bulletin*, vol. 14, no. 353 (7 April 1946), pp. 553-560.

[3] See "Address by the Soviet Representative to the United Nations Atomic Energy Commission (19 June 1946)," in *Documents on Disarmament 1945-1959*, vol. 1, pp. 17-25.

[4] "The Baruch Plan," in ibid., p. 11.

[5] For analysis of the two plans and the negotiations, see Bernhard G. Bechhoefer, *Postwar Negotiations for Arms Control* (Washington, D. C.: Brookings Institution, 1961), pp. 41-82; and Chalmers M. Roberts, *The Nuclear Years: The Arms Race and Arms Control 1945-70* (New York: Praeger, 1970), pp. 9-27.

As a whole, the Baruch Plan was as bold in scope as it was (viewed retrospectively) practically preordained to fail. The United States was the only country with an actual nuclear capability and the experience in producing nuclear weapons. If national military programs in this area were effectively halted, as was called for under the plan, the United States would always retain at least a technological military "edge." This had to be a strong disadvantage from the standpoint of the Soviet Union, which was still in the process of basic research and development on atomic weaponry.[6]

Moreover, "internationalization" of atomic energy programs and prohibitions on any weapon developments meant not only limiting the prerogatives of national sovereignty but also, if safeguards were to be reliable, opening one's borders and doors widely. While such prospects, at least in their general outline, were apparently entertained with no great difficulty in Washington, they would receive quite another reception in Moscow. Taking into account these and other factors, especially the growing differences and general distrust between the United States and the Soviet Union, it comes as no real surprise that the Baruch Plan did not get off the ground.

The reaction of the United States to the lack of success of the Baruch Plan was the inward-looking Atomic Energy Act of 1946. It set up an Atomic Energy Commission under civilian control, provided for government ownership of all fissionable materials and related production facilities, and placed a veil of secrecy over information pertaining to industrial or nonmilitary research and development. The act also precluded any exchanges of information with other countries, including the two close allies which had cooperated in the Manhattan Project.

The Atoms for Peace Program

If the Charybdis of relatively uncontrolled development and dissemination of atomic energy for civil purposes was all too obviously untenable (which it was), the Scylla of excessive controls could not over time be ignored. Would not a U.S. policy of strict secrecy and no information exchanges on potential peaceful benefits serve to encourage indigenous nuclear programs in other countries, programs which

[6] In all fairness, one might also suspect a lack of government-wide wholehearted commitment in Washington to tying control of peaceful applications of atomic energy so absolutely to nuclear disarmament, given the United States edge in the development of nuclear weapons and the estimated Soviet edge in manpower and proximity to western Europe and Asia.

would not necessarily be subject to any international controls or understandings? Would not such a trend decrease whatever chances existed for the establishment of some international safeguards and increase the chances for growing suspicion, instability, and proliferation of nuclear weapons?

Coming closer to home, would not continuing supply of "indispensable quantities of source material . . . from foreign markets" rest on increasingly uneasy grounds so long as the United States continued to foreclose cooperation and the exchange of detailed information in peaceful-uses programs with friendly countries? [7] In addition to these security concerns, did not the political, economic, and scientific interests of the United States argue against maintaining some of the 1946 restrictions, or argue for pursuing cooperative relationships, under safeguards, with selected other countries? So ran the reasoning in Washington early in the 1950s.

President Eisenhower presented the United States "atoms for peace" proposal to the United Nations in December 1953.[8] A comment by Secretary of State Dulles, during his testimony before the Joint Committee on Atomic Energy in 1954, summarized a prevailing view. Knowledge in the atomic energy field was growing "in so much of the world" that the United States could not "effectively dam . . . the flow of information, and if we try to do it we will only dam our own influence and others will move into the field with the bargaining power that that involves." [9]

Legislation to implement "atoms for peace" was contained in the Atomic Energy Act of 1954. It provided the framework and encouragement for participation by American industry in the field of atomic energy. It also established the bases for the promotion of civil uses of atomic energy with other countries to the degree permitted by considerations of national security and defense. The groundwork was laid for the United States to join in cooperative bilateral agreements and, three years later, in the newly founded International Atomic Energy Agency (IAEA). Part of the veil of secrecy necessarily began to be lifted and much—too much, many would maintain to-

---

[7] See John Foster Dulles, "Amending the Atomic Energy Act," statement before the Joint Committee on Atomic Energy (3 June 1954), in *Department of State Bulletin*, vol. 30, no. 781 (14 June 1954), pp. 926-928.

[8] See Dwight D. Eisenhower, "Atomic Power for Peace," address before the United Nations General Assembly (8 December 1953), in ibid., vol. 29, no. 756 (21 December 1953), pp. 847-851.

[9] Testimony of John Foster Dulles, U.S. Congress, Joint Committee on Atomic Energy, *Hearings on S. 3323 and H.R. 8862 to Amend the Atomic Energy Act of 1946*, Part II, 83rd Cong., 2d sess. (1954), pp. 701-702.

day—information was made available. The pendulum had definitely swung in the direction of active promotion of the peaceful atom.

The idea of effective safeguards and controls was not abandoned in the wake of the "atoms for peace" programs. Bilateral or multilateral agreements on the civil uses of atomic energy contained legal provisions for safeguards rights, including monitoring and inspection, and "peaceful uses only" pledges by countries receiving nuclear materials, equipment, or assistance from the United States. Each cooperative agreement proposed by the AEC required presidential approval, as well as a presidential determination that it would "promote and . . . not constitute an unreasonable risk to the common defense and security," and the support or, at least, acquiescence of the Joint Congressional Committee on Atomic Energy.[10] The AEC's licensing procedures for private activities in the United States were to ensure adequate domestic controls and safeguards.

But, while the idea of safeguards and protective requirements was by no means forsaken, it now took a backseat to the promotion of civil uses of atomic energy domestically and internationally. In the words of Willrich and Taylor, "Atoms for Peace . . . signaled a major reordering of priorities. Prior to 1953, international control came first and peaceful nuclear development second. Thereafter, development came first and international inspection and control second, if at all."[11]

The basic shift in favor of development faced little serious, concerted, or effective challenge for well over a decade. The United States, and after 1957 the IAEA, concentrated on promoting peaceful applications of the atom in medicine, agriculture, and industry. This promotion was particularly focused on research and power reactors which, in and of themselves or when not combined with a reprocessing capability, presented the least security risk of any part of the nuclear fuel cycle—so long as they were subject to safeguards and the technology itself was protected.[12] The list of research and research-power agreements expanded, reaching eighteen agreements

[10] See Atomic Energy Act of 1954, Chap. 11, sec. 123.

[11] Willrich and Taylor, *Nuclear Theft*, p. 180.

[12] It should be noted, in this regard, that different types of reactors present different safeguard considerations. Light-water reactors (LWRs) are generally easier to safeguard than heavy-water reactors (HWRs). The former require periodic shutdown for refueling and are fueled with enriched uranium (the supply of which is limited to countries with enrichment capabilities), whereas the latter have on-line reloading and natural uranium as fuel. Standard safeguards on both types, however, should detect any significant, unusual operations.

by 1960; [13] and other supplier countries such as Great Britain, Canada, and France entered the picture.

By the late 1950s and early 1960s, interest in the problem of proliferation of nuclear weapons was growing steadily, focusing mainly from an arms control (in contrast to technology control, for example) point of view. Nonproliferation concerns inhered in the cutoff of fissionable materials for military programs and the comprehensive test ban proposals, and in the 1963 Limited Test Ban Treaty. For a concern about the spread of nuclear weapon capabilities, however, such approaches were at best indirect. A United Nations resolution in 1961, calling for a nonproliferation treaty, was not. Negotiations on such a treaty began the following year at the Eighteen Nation Disarmament Conference, now the Conference of the Committee on Disarmament, at Geneva. This period also marked the detonation of the first French nuclear device in 1960, and then in 1964 the People's Republic of China entered the so-called nuclear club.

The Nuclear Nonproliferation Treaty (NPT)

The cause of nuclear nonproliferation was boosted substantially in the late 1960s with the advancing negotiations and conclusion of the NPT. This landmark agreement, now adhered to by over ninety countries, reflected the widespread concern that nuclear proliferation "would seriously enhance the danger of nuclear war" with all its devastating consequences for those affected and indeed, on any large scale, for the world.

To reduce this danger, two different classes of parties to the NPT, nuclear-weapon states and non-nuclear-weapon states, undertook several key international legal obligations. Nuclear-weapon states [14]

[13] By 1960, the list included research agreements with Denmark, Greece, Iran, Ireland, Israel, the Republic of China (Taiwan), South Korea, Thailand, Turkey, the United Kingdom, and South Vietnam; research-power agreements with Australia, Canada, Italy, South Africa, and Spain; a joint power program with the European Atomic Energy Community (EURATOM); and an agreement for supply of materials with the IAEA. By 1970, the United States had concluded agreements for cooperation (mostly research-power agreements today) with over thirty countries. Added to the above list were Argentina, Austria, Brazil, Colombia, Finland, India, Indonesia, Japan, Norway, the Philippines, Portugal, Sweden, Switzerland, and Venezuela.

[14] Article IX of the NPT declares: "For the purposes of this Treaty, a nuclear-weapon State is one which has manufactured and exploded a nuclear weapon or other nuclear explosive device prior to January 1, 1967."

Of the five countries in this category, the United States, the Soviet Union, and Great Britain are parties to the NPT; France and the People's Republic of

party to the treaty agreed not to transfer nuclear explosive devices to any other country (nuclear or non-nuclear) and not to facilitate, in any way, the acquisition of such devices by any non-nuclear-weapon state. Non-nuclear parties to the treaty[15] agreed not to construct or otherwise acquire nuclear explosives for military or nonmilitary purposes and to allow IAEA safeguards on all their nuclear activities, whether indigenous or based on materials and equipment imported from other countries. All parties assumed an obligation to require the application of IAEA safeguards on any exports of special fissionable material and related processing or production equipment and materials.

The treaty laid the groundwork for the establishment of an expanded IAEA safeguards system and stimulated further safeguards efforts among nuclear supplier countries belonging to the IAEA. Through its combination of a broad prohibition against the spread of nuclear explosives with specific safeguards obligations, the NPT represents a major milestone in the nonproliferation effort. Common political-security denominators of an emerging consensus mixed with some political bargains and compromises, and were forged into a formidable and valuable legal structure.

But the NPT did not, nor did it pretend to, provide all the answers to the many facets of the nuclear nonproliferation problem. An assumption of universal adherence and permanent observance was as unwarranted in the late 1960s as it is today. In addition to the problem of subnational threats and a few outstanding questions of detail relating to the NPT itself, the need to pursue additional approaches toward containing and resolving proliferation issues derives from two basic considerations.

First, however desirable from a nonproliferation standpoint, universal or near universal adherence to the NPT is not likely in the foreseeable future. Over fifty countries—including India, Pakistan, Israel, Egypt, Brazil, Argentina, South Africa, Turkey, Portugal, Spain, France, and the People's Republic of China—have not ratified the treaty. (Japan has also not ratified, but has signed and is moving

China are not, and have shown no interest in becoming parties to the treaty. France has, however, frequently stated its intention to act in a manner consistent with the treaty's objectives. The PRC has denounced the treaty, but there is no evidence that Peking has acted or intends to act contrary to the prohibitions on weapon transfer or assistance. India cannot be included within the NPT's definition of a nuclear-weapon state because of the cutoff date set forth in the treaty. (For more on the position of India, see Chapter 2.)

[15] Non-nuclear-weapon parties to the NPT include all non-nuclear members of NATO except Turkey and Portugal, all non-nuclear members of the Warsaw Pact, and several Arab, Asian, Latin American and African countries. See appendix for complete list of parties to the NPT as of mid-1975.

toward ratification.) Significantly, five of the above nonparties to the treaty are not only nuclear energy users but also present or potential suppliers of nuclear materials and equipment. The nonparty element might change for the better in the years ahead. West Germany, Italy, Belgium, the Netherlands, South Korea, Libya, and a few other countries ratified the NPT in the first half of 1975. In the meantime, however, universal adherence is not expected and there are technology control, safeguards, and other issues to be dealt with in the current given context.

Second, even if the NPT is more widely embraced, the treaty cannot surely guarantee the future even though it can help mold it. The NPT stipulates that all parties undertake "to facilitate, and have the right to participate in, the fullest possible exchange of equipment, materials, and scientific and technological information for the peaceful uses of nuclear energy." Nevertheless, simple prudence has demanded from the outset that criteria other than NPT membership had to be brought to bear on questions of nuclear cooperation—criteria such as the sensitivity, security implications, or potential military significance of the nuclear materials or technology under consideration for transfer, and the political stability or instability in the country or region involved.

The need to apply additional criteria reflects the fact that NPT membership alone cannot dissolve concern about the accumulation of the preconditions for a nuclear explosive capability in certain countries and regions. Security perspectives could change over time and, however undesirable from a nonproliferation standpoint, so could a country's allegiance to the NPT. Article X of the treaty provides a standard mechanism for a party to withdraw from its obligations on three months notice if it determines that "extraordinary events" have placed its "supreme interests" in jeopardy. It cannot confidently be assumed that the NPT regime will remain stable over time.

Inherent limitations on what the NPT itself can accomplish do not diminish the interest of the United States in actively supporting and—when and where possible—strengthening the political-legal regime of the treaty. In this regard, however, certain constraints on American foreign policy will remain operative. First the pool of U.S. influence is limited. The expenditure of influence directly on NPT issues, generally or specifically vis-à-vis nonparties, must be weighed within the context of a wide range of issues. Spending more on the NPT issue could mean adversely affecting or taking away from other matters, even other antiproliferation efforts, deemed equally or more important. Second, the ability of the United States to influence is also limited. There are few, if any, countries not party to the NPT where additional U.S. influence or pressure would constitute a major, much less a decisive, contribution to a decision to adhere to the treaty; and certainly none where this could be attempted without risk or costs.

These points can take on a darker or lighter shade in the analysis of individual countries not party to the NPT. In some, perhaps more scope exists for U.S. influence than has been exercised to date; in others, there may be less scope than is commonly assumed. Some disagreement will probably persist in Washington on how, when, and where the expenditure of influence can be productive.[16] Whatever the political judgments on specific countries and methods, the general constraints cannot be ignored without cost.

Recognition of the limited reach of the NPT does not detract from its special value. Nor does awareness of limitations on Washington's influence argue against its being exercised in general. Support for the treaty and for wider adherence to it, and the fulfillment of U.S. obligations under it will remain important parts of United States policy on nuclear nonproliferation. But the NPT constitutes only one part of that policy and, in the progression of events since its negotiation, not necessarily the most consequential component today.

[16] Some proponents of the treaty have persistently argued that the Nixon administration could and should have done more to secure NPT ratification by key nonparties. That assertion will remain open to question, as will a "strong and more open pressure" approach to other countries. The issue has been altered in 1975 with NPT ratification by several countries and with Japan's moving closer to ratification. Since some disagreement may well continue, it is worth noting that the issue does not really concern support for the substance of the treaty or for wider adherence, but rather how much and what kind of influence the United States should bring to bear in attempting to move nonparties in the direction of ratification.

2 PERSPECTIVES ON AN EXPANDING PARADOX

Do existing controls—international safeguards, export control criteria, export review processes, national physical security measures, and materials accounting procedures—provide adequate assurances against the risk of national diversion or takeover for military purposes, and the risk of subnational diversion, theft, or sabotage?

In the early 1970s, pursuant to a presidential directive, issues pertaining to this query began to be assessed by the government agencies with longstanding interests in one or more of its aspects (primarily the AEC but also the Department of State, the Arms Control and Disarmament Agency, and the Department of Defense). The implications of rapidly expanding nuclear energy industries, both at home and extending into several dozens of countries, raised serious doubts about the sufficiency of established protective measures. Looking at a sketch of these implications (focusing first on the processes from which nuclear weapon-usable materials can be derived and, then, on the materials themselves and associated risks), the question remains very much alive today, especially on the international level.

Nuclear Energy Outlook and Risks

Uranium Enrichment. Natural uranium must be enriched to fuel light-water reactors (LWRs), the predominant type of operating reactor in the United States and abroad, and a few other types of reactors. High-temperature gas-cooled reactors (HTGRs), now in their demonstration stage, will require enriched uranium fuel. Only heavy-water reactors (HWRs) and one kind of gas-cooled reactor (the GCR) use natural uranium fuel. The growth of nuclear power industries, expected to reach hundreds of enriched uranium-fueled reactors over the next two decades, will thus demand continued expansion of uranium enrichment or isotope separation capabilities.

The United States, the Soviet Union, the People's Republic of China, France, and the United Kingdom currently have uranium enrichment facilities based on diffusion technology; France plans to build another diffusion plant for commercial purposes; and Canada has an interest in constructing one. The controlled spread or sharing of diffusion technology, under safeguards, has several advantages in terms of safeguards and nonproliferation objectives. Indeed, in 1971, the United States offered to sell its advanced diffusion technology to allies for use in a multinational, safeguarded facility. These plants are economical only on comparatively large scales and very expensive both to build and (because of high electricity requirements) to operate. Moreover, a diffusion facility built for low enrichment could not be converted to the production of high-enriched, weapon-grade uranium without substantial alterations which should readily be detected by safeguards. If the technology were acquired or developed by third countries, few could afford a diffusion plant, and it is not likely that any could be concealed.

A major concern from the standpoint of nuclear proliferation is the development and possible spread of advanced uranium enrichment technologies based on centrifuge or laser methods of isotope separation.[1] Programs for developing one or both of these technologies are going forward in several countries, including the United States, the Soviet Union, West Germany, Great Britain, the Netherlands, and Japan. Recognizing that the key problem is one of controlling the spread of enrichment capabilities generally, regardless of the technological process involved, the possible implications of these advanced techniques nevertheless raise special concerns.

---

[1] Another method of isotope separation being developed is the so-called nozzle or stationary-wall centrifuge process. Although the technology to be used in the South African uranium enrichment plant has not been confirmed, indications are that it is based on this process rather than the other technologies mentioned above. A West German group (Becker) has advanced the nozzle method, and it is the enrichment technique involved in the 1975 West Germany-Brazil nuclear energy agreement.

While it remains debatable whether or not a facility based on nozzle technology could be converted to the production of high-enriched uranium more easily than could a diffusion plant, the estimated size and power requirements for nozzle plants are similar to those for diffusion plants. Thus the technology does not appear to present the same problems as centrifuge and laser methods of isotope separation. Nevertheless, nozzle technology must still be considered part of the key problem noted in the text—how to control the spread of enrichment capabilities generally.

The development of laser methods of uranium enrichment, principally in the United States, is still in its early stages with the outcome uncertain—but with expectations that the process might substantially reduce enrichment costs. Although much cannot yet be said about the possible parameters of a laser facility, should the technology pan out, it appears that such a plant could be significantly smaller in scale and have lower power requirements than one based on diffusion technology. In these respects, laser technology presents potential safeguards and proliferation problems similar to those discussed below for the centrifuge. These potential problems are further compounded in the case of laser technology by the fact that the product of the process would be high-enriched, weapon-grade uranium, whereas centrifuge facilities can be set up for either low or high enrichment.

Centrifuge development, in contrast to laser technology, is well on its way in the United States and western Europe. The centrifuge process holds out the promise, *inter alia,* of considerably smaller, commercially competitive facilities that would be much less expensive to build than diffusion plants (not necessarily per unit of output but because of the differences in size), and much less expensive to operate (because of lower electricity requirements). The possession of centrifuge technology and facilities by those countries known to be actively engaged in its development presents no direct, significant, foreseeable risks to nuclear nonproliferation objectives. But the promise of centrifuge technology contains obvious potential security risks. A smaller and less costly plant could be afforded by more countries (thus facilitating spread of enrichment capabilities), and could perhaps be concealed without great difficulty.

An additional concern has been that a centrifuge facility set up for the production of low-enriched or nonweapon-grade uranium could be altered to produce high-enriched or weapon-grade stuff more rapidly, and with considerably less difficulty and less obvious alterations, than would be involved in converting a diffusion plant built for low enrichment. This does not mean that centrifuge facilities, not to mention the possibility of facilities based on laser technology, are inherently "unsafeguardable." (They are not, although it remains in large part to be seen if technical requirements can be devised that are economically and politically acceptable to countries interested in developing the process.) But it does imply a shorter lead time to the production of weapon-grade uranium, in the event facilities were taken over for such purposes, and perhaps increased problems for safeguards against covert diversion.

Chemical Reprocessing. Nuclear reactors produce not only electricity but also plutonium. The security risks associated with this product of nuclear power generation could be eliminated, for all practical purposes, by leaving it mixed in unreprocessed spent reactor fuel. But no reprocessing to separate plutonium would also mean foregoing both plutonium recycle (which could meet 10 to 20 percent of the fuel requirements for LWRs in the years ahead) and breeder reactors (which use plutonium fuel). While plutonium recycle and fast-breeder reactors (FBRs) have been the subject of debate, largely because of problems linked to plutonium, the current forecast for the growth of nuclear power industries includes the expansion and spread of reprocessing technology, equipment, and facilities. Argentina, Belgium, the People's Republic of China, France, West Germany, Great Britain, India, Italy, Japan, the Soviet Union, Spain, and the United States have pilot plant or commercial reprocessing facilities in operation or under construction. Canada, Sweden, and Brazil have plans for developing reprocessing capabilities.

It is apparent from the foregoing discussion that uranium enrichment and chemical reprocessing are the central considerations in the context of technology control to reduce the security risks of the peaceful atom. Where and how, and under what conditions, should the spread of these capabilities take place? Should the spread be carefully controlled? If so, how? Do the nonreplication provisions in standard commercial licensing procedures really provide much assurance over time? In the context of measures against diversion, national or subnational, key control points are enrichment and reprocessing facilities, along with fuel fabrication and conversion facilities for high-enriched uranium or plutonium. Two other key areas for control are, of course, the weapon-usable materials in storage and in transit prior to their being irradiated in a reactor and, especially, prior to their being chemically coated in the fuel fabrication process.

High-Enriched Uranium. Significant quantities of high-enriched weapon-grade uranium have been used over the years and transferred to other countries, under safeguards, for research reactor programs. The risks associated with high-enriched uranium, and the issues of safeguards and controls generally, were brought into somewhat sharper focus in the early 1970s by the advent of the high-temperature gas-cooled reactor (HTGR) in the United States and the forecast of sales and development abroad. Neither natural nor low-enriched uranium (used to fuel current operating reactors) is usable for nuclear explosive purposes. HTGRs, on the other hand, will require the production of large quantities of high-enriched, weapon-grade uranium.

High-enriched uranium and plutonium, as will be noted, present much the same risks and require the same safeguards and transfer controls. The two differ, however, in a few notable respects. Although the estimated quantities of high-enriched uranium are not at all on the same order of magnitude as the estimated amounts of plutonium, the physical properties of the former make it easier to handle and work with in the construction of a nuclear device. On the other hand, these same physical properties mean that high-enriched uranium does not represent a similar potential contamination-health threat.

Plutonium. It is estimated that amounts of plutonium in the private sector, available from LWRs in the United States, will rise from a cumulative total of 730 kilograms in the early 1970s to over 150,000 kilograms by 1985. Assuming that FBRs come on line in the 1990s, as forecast, quantities of plutonium in use, storage, and transit in the private sector are expected to exceed 1,500,000 kilograms or 1,500 metric tons by the year 2000. Figures 1 and 2 graphically illustrate the expected rapid rise of nuclear power capacity and plutonium recovery in the United States.

Depending upon the assumptions one makes, the nuclear power capacity of all foreign countries combined is expected to be at least 50 percent greater than, and perhaps more than double, that of the United States by the year 2000.[2] Thus the estimated total amount of plutonium to be available from power reactors in other countries, taken together, will range by that year from 2,300,000 to well over 3,000,000 kilograms. According to the number of power reactors operating, ordered, announced, and planned, nuclear energy programs abroad will be concentrated in eleven countries: Japan, Great Britain,

[2] For analysis of estimated ranges in nuclear power capacity in the United States and abroad, see U.S. Atomic Energy Commission, *Nuclear Power Growth 1974-2000*, AEC/WASH 1139, February 1974, pp. 1-23 and 33-37; and "Nuclear Fuel Cycles: 1973-80," "Nuclear Power Scenarios: 1980-2000," and "Foreign Nuclear Power: Reactor Types and Forecasts," in Willrich and Taylor, *Nuclear Theft*, pp. 29-58, 59-76, and 193-201 respectively.

According to AEC estimates in February 1974, the United States has 42 operating power reactors, 147 on order, 28 announced, and 18 planned, for a total of 235 reactors. The figures for foreign countries, taken together, include 86 operating, 134 ordered, 25 announced, and 139 planned, for a total of 384 reactors. These figures do not take into account more recent announcements, such as those by Brazil and Iran, or changes in plans as a result of the energy crisis. See *Nuclear Power Growth 1974-2000*, pp. 67-74.

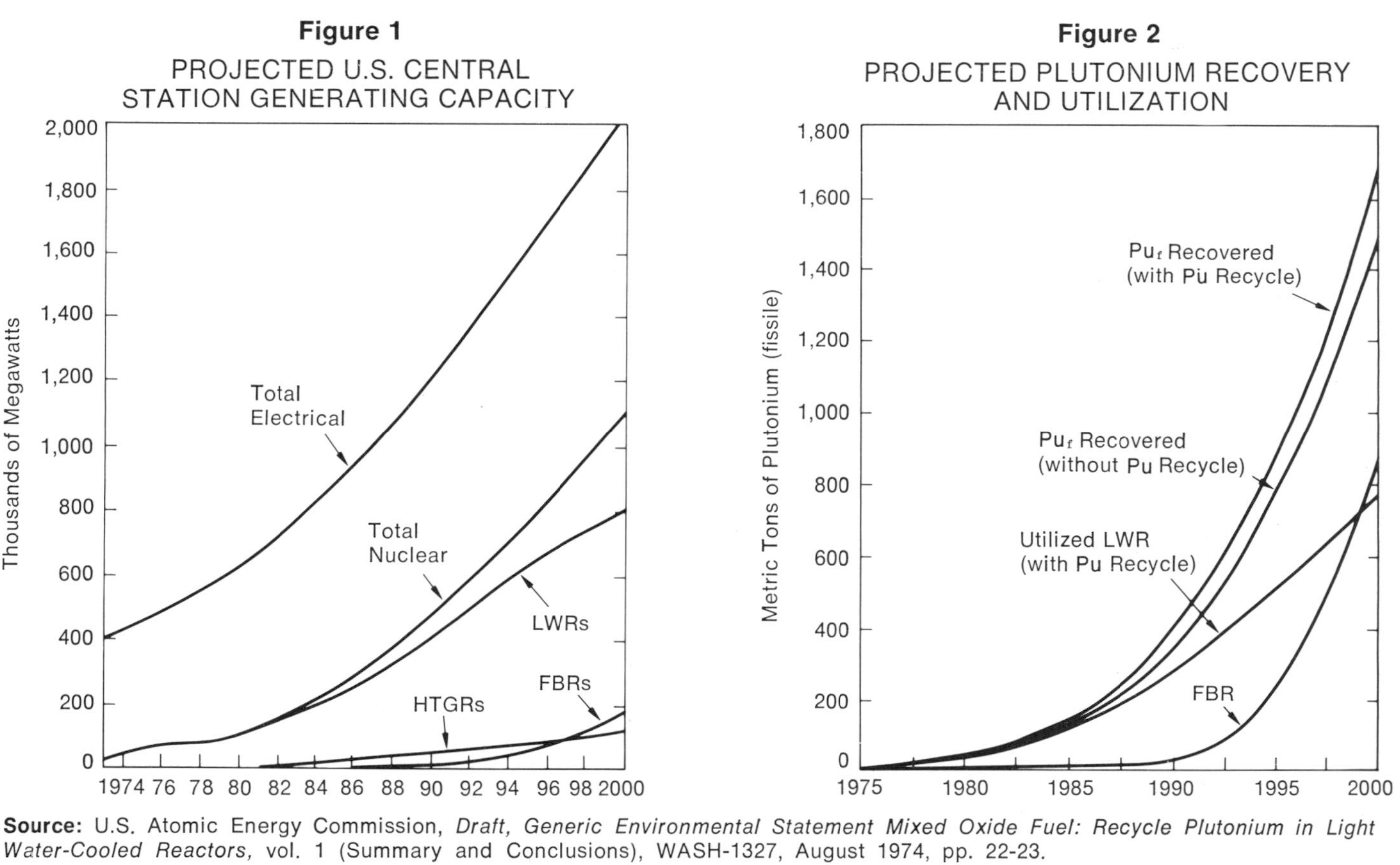

Figure 1
PROJECTED U.S. CENTRAL STATION GENERATING CAPACITY

Figure 2
PROJECTED PLUTONIUM RECOVERY AND UTILIZATION

Source: U.S. Atomic Energy Commission, *Draft, Generic Environmental Statement Mixed Oxide Fuel: Recycle Plutonium in Light Water-Cooled Reactors,* vol. 1 (Summary and Conclusions), WASH-1327, August 1974, pp. 22-23.

West Germany, the Soviet Union, Canada, France, Sweden, Spain, India, Switzerland, and Iran.[3]

By the mid-1980s, however, it has been estimated that a minimum of thirty-six other countries will have operating nuclear power reactors.[4] A single large reactor, on the scale of 1,000 megawatts electrical, produces a few hundred kilograms of plutonium annually or enough for twenty-five or more nuclear explosives.[5] The annual spent fuel from one medium-size reactor for generating electricity contains enough plutonium for some nuclear devices; and, not to be forgotten, significant quantities of plutonium can be accumulated, over time, from a small research reactor.

Associated Risks. A few kilograms of plutonium and not much more of high-enriched uranium would be enough for the construction of a nuclear device. Because of its high radiotoxicity and the consequent radiobiological hazards from any exposure, plutonium is quite difficult to work with and requires a number of strict precautions. Working with high-enriched uranium is, as noted, less difficult. At any rate, adequate precautions can be taken for either material by a rea-

[3] Based essentially on pre-energy crisis estimates, with the exception of the figure for Iran, the number of operating, ordered, announced, and planned reactors in these countries can be broken down roughly as follows: Japan (70); Great Britain (48); West Germany (38); Canada and the Soviet Union (25); France and Sweden (20); Spain (14); India (11); Switzerland (10); and Iran (10+). See ibid.

[4] Based essentially on pre-energy crisis estimates, the countries in the range of five-to-nine operating, ordered, announced, or planned reactors include Argentina, Belgium, Brazil (perhaps more in light of the 1975 agreement with West Germany), Bulgaria, Czechoslovakia, Finland, East Germany, Italy, and Taiwan. Those in the less-than-five category (mostly in the one-to-three range) include Australia, Austria, Bangladesh, Chile, Egypt, Greece, Hong Kong, Hungary, Ireland, Israel, Jamaica, South Korea, Luxembourg, Mexico, the Netherlands, New Zealand, Norway, Pakistan, the Philippines, Poland, Portugal, Romania, Singapore, South Africa, Thailand, Turkey, and Yugoslavia. See ibid.

[5] A few distinctions deserve mention in this regard. Plutonium from the spent fuel of an LWR or HWR, operated economically for generating electricity, is weapon-usable. But it is more "contaminated" (with a higher percentage of plutonium-240 content) and, therefore, less reliable for weapons purposes than plutonium produced in a manner specifically for weapons or than that which will be available from FBRs when they come on line. A reactor could be operated, although uneconomically from the standpoint of efficient electricity generation, to produce better-grade plutonium. It is unlikely, however, that one subject to safeguards could be thus operated without detection.

The importance of higher reliability in weapon yield depends upon the scenarios postulated for a nation going nuclear. Higher reliability would be essential in some scenarios, and the preferred alternative for any nation deciding to acquire nuclear devices. But the degree of reliability may not be an important distinction for a small covert national program, or for national takeover in desperation or crisis scenarios. The degree of reliability is probably not a distinction at all for subnational threat scenarios.

sonably competent technical individual or group. Also, published information on nuclear research and development has greatly lowered the technical barriers to nuclear device design and construction, whether by national authorities or by dissident or terrorist groups.

In quantities of much less than one kilogram, plutonium could be attached to a few pounds of readily available high explosives to present a very serious contamination-health threat in populated areas: "Inhalation of particles the size of specks of dust and weighing a total of some ten millionths of a gram is likely to cause lung cancer. A few thousandths of a gram of small particles of plutonium (taken together, about the size of a pinhead), if inhaled, can cause death from fibrosis of the lungs within a few weeks or less." [6] Compounding this potential threat is the fact that decontamination of an affected area would, at best, be a complex, costly, and time-consuming task. Since the contamination-health threat avoids the intricacies, difficulties, and dangers of nuclear device construction, it is particularly relevant when considering risks at the subnational level.

In light of the amount of materials required for a nuclear device, the not uncommon statement that the projected amount of plutonium available from nuclear power reactors represents several thousand potential devices is essentially correct. But, taken by itself, it is somewhat misleading and misses the core of the problem. By far the greatest amounts of plutonium will be either (1) in nuclear-weapon countries, which present an arms control problem but not a diversion-proliferation risk as considered in this essay, or (2) in non-nuclear-weapon countries where there are strong disincentives, or no significant incentives, for developing nuclear weapon capabilities. The central problem is that potential threats do not necessarily require much plutonium or high-enriched uranium, and very large amounts of these materials (particularly plutonium) in peaceful programs can increase the chances for its acquisition for noncivil purposes.

At the national level of risk, a relatively small nuclear weapon capability, no matter whether a country labels it for peaceful purposes and conducts no overt military production program, could have significant regional, if not necessarily global or strategic, consequences. The distance between the two realms of consequences can be quite short—as is highlighted, for example, by the situation in the Middle East. The necessary amount of plutonium or high-enriched uranium would be correspondingly small if a country were to perceive that a few nuclear devices, with or without highly sophisticated delivery systems, would be in its interest on security or other grounds.

[6] Willrich and Taylor, *Nuclear Theft*, p. 13.

Again, either a reprocessing or an enrichment capability would provide a path to the acquisition of weapon-usable materials. Such materials might also be acquired from another country or, not inconceivably, from a black-market operation.

At the subnational level of risk, much smaller quantities of plutonium or high-enriched uranium take on significance. One crude nuclear device—the basic principles for the construction of which are available in published materials—could present a grave threat in the hands of a seriously disaffected individual or terrorist group. So could much less than one kilogram of plutonium attached to high explosives.

Considering the broad outlines of the nuclear energy outlook and the associated risks, it is easy to understand why important doubts about the adequacy of existing safeguards and controls were gaining a wide hearing, both inside and outside the government, by the early 1970s. Washington began to put its own house in order by substantially upgrading requirements for physical security measures and materials accounting procedures in the private sector, and by establishing more stringent guidelines and conditions for any export from the United States of significant quantities of high-enriched uranium or plutonium. Long before the Indian detonation of May 1974, the interested government agencies had been in the process of hammering out various steps the United States could consider taking with other countries to curb the security risks of the peaceful atom.

Controversy pushes issues forward and up. In the late spring and early summer of 1974, the U.S. government shifted gears in its planning. The Indian test and then Washington's proposal to provide nuclear energy assistance to Egypt and Israel dramatized the issues, intensified concerns in nongovernmental circles about nuclear proliferation and subnational threats, and lent a renewed sense of urgency within the government to the need for resolving outstanding issues to the degree possible.

A Bang from India

A nuclear test beneath the Rajasthan desert, on 18 May 1974, declared India as the sixth member of the so-called nuclear club. The test did not come as a complete surprise abroad. India had undertaken no international obligations, as contained in the NPT, that would stand in the way of its acquiring nuclear explosives; and international safeguards were not applied either to India's indigenous reprocessing facility or to a natural uranium reactor supplied by Canada in the

1950s.[7] Many intelligence and nuclear affairs experts had generally concluded, in the few years before 1974, that Indian acquisition of nuclear explosives was most likely a matter of "when" rather than "whether." The test came, nonetheless, as a major disappointment to many outside India. The curtain dropped on a decade when the number of nuclear explosive-capable countries had remained at five, and interest intensified in the prospects and problems of nuclear proliferation.

India stressed that its nuclear test and continuing program were dedicated to the development of nuclear explosive technology for peaceful or nonmilitary applications—such as use in mining or oil recovery—and not designed for weapons or military purposes. Yet the technology of nuclear devices for civil purposes remains indistinguishable from that for weapons. New Delhi no doubt recognized the unavoidable ambiguity in its test, and this ambiguity could actually appear desirable in the eyes of India's leadership if security factors played some role in the decision-making process along with political, prestige, technical, or economic considerations. At any rate, the "peaceful purposes" label would appear to many beholders as a distinction without much, if any, difference.[8] Even if one granted complete Indian sincerity in the present declared policy against acquisition of nuclear weapons, a question mark would naturally hover over the future. What would happen to the "peaceful purposes" distinction in time when, according to 1974 polling reports, public opinion in India strongly favored the Rajasthan test and the idea of India acquiring a military nuclear capability?

The responses of other countries to the test were probably about what New Delhi had calculated, with the possible exception of the

---

[7] India obtained plutonium for its nuclear device from the reactor supplied by Canada in the 1950s, prior to the establishment of the IAEA safeguards system and at a time when there was probably very little, if any, outside concern about an Indian nuclear outlook. The United States, in years past, supplied heavy water for the reactor.

[8] As an example of the "peaceful purposes" label making little practical difference, the Agency for the Prohibition of Nuclear Weapons in Latin America (OPANAL) has not only referred to India as one of the "nuclear powers," but also approached India to gain its adherence to Protocol II of the Latin American Nuclear Free Zone Treaty. Protocol II contains certain obligations for nuclear-weapon countries to respect the treaty and not to use, or threaten to use, nuclear weapons against the parties. The United States, Great Britain, France, and the People's Republic of China have adhered to the protocol, while the Soviet Union has not. New Delhi has apparently claimed that the protocol for nuclear-weapon countries is not applicable to India. For reference to India as a nuclear power, see "OPANAL Report on the Implementation of the Treaty of Tlatelolco," document submitted to the NPT Review Conference, NPT/CONF 9, 24 February 1975, p. 5.

mildness of the public statement of the United States. Pakistan, a neighbor uneasy with reason, sharply criticized India's action and stressed that clearer security guarantees by the major powers against potential nuclear blackmail or attack were necessary. Privately, Pakistan may well have looked toward technically upgrading its own peaceful nuclear programs which could, in time, provide it with a capability to develop nuclear explosives. Reports in 1975 indicated that Pakistan was seeking reprocessing facilities from France even though the former had, as of January 1974, only one small nuclear reactor operating and another one on order.[9]

Canada also found no reason to soften its opposition to the Indian test, and quickly suspended all its nuclear assistance to India's civil nuclear programs. Besides being one of the most vigorous and consistent supporters of nuclear nonproliferation, Ottawa was confronted with what it could only view as New Delhi's violation of what had been thought to be a mutually understood condition—that nuclear materials and equipment supplied by Canada, and materials derived therefrom, would not be used for any nuclear explosive purposes. New Delhi's interpretation of the supply conditions obviously differed from Ottawa's. (The discussions between the two capitals on the resumption of assistance, undertaken in the aftermath of its suspension, will presumably progress or falter on the basis of Ottawa's understanding.)

Clear expressions of disapproval came from several more capitals. Still others, particularly a few third world capitals, found a fair amount of satisfaction in the Indian detonation. The public responses from the People's Republic of China and the two nuclear superpowers, on the other hand, were marked by conspicuous mildness. Peking may well be determined to keep a watchful eye on Indian nuclear developments and, equally important from Peking's view, delivery capabilities; but it played down the test issue publicly.

Moscow's general commitment to nuclear nonproliferation gave way—as is often the case when any country's pursuit of a generalized preference butts into more specific interests—primarily before its desire not to disturb the development of a closer cooperative relation-

[9] Prime Minister Bhutto reportedly expressed a willingness, during his discussions in Washington, to accept more stringent safeguards (such as those proposed by the United States with respect to nuclear assistance to Egypt and Israel) on all of Pakistan's peaceful nuclear energy programs regardless of supplier. This willingness, however, was apparently tied into the talks regarding Pakistan's perceived need for increased conventional arms and resumed supply from the United States, and does not seem to extend to refraining from the acquisition of reprocessing capabilities. For report on the Washington meeting, see *Washington Star-News*, 6 February 1975, p. 5.

ship with New Delhi. At a secondary level of importance, Moscow may also have wished not to appear contrary to the idea of peaceful nuclear explosions (PNEs) because of a sizable investment in its own peaceful applications program. Whatever the reasons, Moscow's response highlighted India's affirmation that its program was dedicated to peaceful purposes only.

The considerations regarding how to respond were not totally dissimilar in Washington. Washington had scant interest in any PNEs, but did wish to pursue a continued improvement in relations with New Delhi—relations which had deteriorated substantially during the India-Pakistan war of 1971. The latter interest, punctuated by Secretary Kissinger's planned visit to New Delhi, argued against strongly declared opposition to India's action. For this reason and no doubt for others (such as the weaknesses of the arguments favoring a stronger public stand), the State Department's release on 18 May simply stated that "the United States has always been against nuclear proliferation for the adverse impact it will have on world stability." [10]

Even though it is not known to what degree these and other related factors played a role in determining Secretary Kissinger's preferences and Washington's response, it is worthwhile to note that nonproliferation considerations themselves can pull in different directions. On the one hand, for example, denunciation and follow-on measures, such as the suspension of nuclear assistance and sales, could signal that Washington meant business in its nonproliferation policy. On the other hand, no matter whether such a signal would carry any effective weight anywhere, it would diminish Washington's already quite limited influence and leverage in New Delhi and would possibly detract from other nonproliferation objectives such as controls and safeguards on nuclear transfers.

Washington's immediate public response was followed by other moves in less noisy diplomatic channels. In a June speech before the IAEA Board of Governors, the American ambassador to the IAEA reiterated two U.S. understandings of its atomic energy agreements with other countries: No material or equipment supplied by the United States could be used for any nuclear explosive devices, and the IAEA would verify, among other things, whether or not safeguarded mate-

[10] *New York Times*, 19 May 1974, p. 19. Many observers at home and abroad considered the response overly gentle. Even India may have been somewhat surprised. Yet, as subsequently suggested by one keen student of proliferation problems, even an attitude of "boredom and indifference" to the Indian test could support nonproliferation objectives if it contributed to an atmosphere of assigning less "attention, glory, or prestige to entry into the 'nuclear club.'" See George H. Quester, "Can Proliferation Now Be Stopped?" *Foreign Affairs*, vol. 53 (October 1974), p. 80.

rial was being used for such purposes. These understandings had not been explicitly provided for, over the years, either in U.S. agreements for cooperation or in the IAEA's safeguards agreements. Nevertheless, Washington considered them fundamental and warned that they had to be respected if American cooperation in civil uses of nuclear energy were to continue.

Prior to shipping nuclear reactor materials already ordered by India, Washington sought and presumably received some assurance on these understandings. This assurance could have little effect on India's current nuclear explosive program since neither its source of plutonium nor its reprocessing capability is dependent on supply of materials or equipment from the United States. But it could have a productive political effect from Washington's point of view, and reinforce Ottawa's longstanding position against the use of plutonium from the reactor supplied by Canada for any explosive purposes.

The subject of the spread of nuclear technology and its implications for nuclear proliferation was discussed during Secretary Kissinger's visit to New Delhi at the end of October 1974. The result included a reaffirmation of "India's policy not to develop nuclear weapons," as well as a statement of "mutual recognition of the need of putting nuclear technology to constructive use . . . and of ensuring that nuclear energy does not contribute to any proliferation of nuclear weapons."[11] This result left untouched India's "peaceful nuclear explosive" program, its rationale, and its apparent contradiction to a policy of nonproliferation of nuclear weapons. There was also no indication that the idea of India's accepting some international observation on its program, in order to provide assurances to others regarding its nature and scope, had been mentioned. But no one expected New Delhi to put its nuclear genie back in a bottle, and the communiqué did indicate its willingness to avoid actions—presumably actions such as the transfer to third countries of nuclear explosive technology or devices—that could contribute to the further proliferation of nuclear explosive capabilities.

A Bid to Egypt and Israel

Less than a month after India's blast, President Nixon announced from Cairo that the United States would negotiate an agreement for cooperation with Egypt to provide nuclear reactors and fuel. Three

[11] Joint Communiqué, issued at the conclusion of Secretary of State Kissinger's visit to New Delhi, Department of State Press Release, no. 449 (30 October 1974), p. 2.

days later, on 17 June, a similar presidential announcement was made with respect to concluding an agreement with Israel. Increasing (but not illegitimate) interest in nuclear power in the Middle East accentuated the security problems associated with the peaceful atom.

Although it was hoped in Washington that the nuclear energy proposals would symbolize the beginnings of a new relationship with Egypt and a more balanced or evenhanded approach to the Middle East, these proposals were not last minute concoctions to curry favor with Cairo. There had been intermittent contact between Egyptian authorities and representatives of American companies concerning the possible sale of reactors for several years prior to the June 1974 statements.[12] For fairly obvious reasons, such as the cool state of relations between Cairo and Washington and the overall Middle East situation, neither Egypt nor the United States actively pursued the matter of nuclear reactor sales until a few months after the disengagement diplomacy of late 1973 and early 1974.

By that time, another consideration had entered the picture—Israeli interest in acquiring nuclear power reactors from the United States. In a sense, this interest facilitated examination of the question of supplying nuclear reactors to Egypt or any other Middle East country. It was evident from the outset of intragovernmental discussions in Washington that certain controls, extending beyond IAEA safeguards and the bilateral arrangements in other agreements for cooperation, should be placed on the supply of nuclear materials to the Middle East. Egypt and Israel might not favor additional controls for themselves, but they might well perceive the desirability of having them placed on each other. If the principle of additional controls were accepted in this case, its application in future similar cases could be made easier. In this light, a set of preconditions for reactor supply to Egypt and Israel were carefully developed by the interested agencies before some consultations on Capitol Hill and the presidential announcements in June. The preconditions were designed to establish a considerably more controlled framework for supply than the standard provisions contained in existing agreements for cooperation.

The proposed sale of reactors and fuel was, nevertheless, bound to give rise to serious questions about the adequacy of international safeguards and other controls (existing and newly suggested) on the peaceful applications of nuclear energy. Egypt possesses neither the

[12] In the absence of a basis on which to call off such preliminary, private contacts, Washington did stress that Egyptian authorities should be fully apprised of U.S. governmental and legal requirements pertaining to any supply of reactors or fuel. This meant the conclusion of an intergovernmental agreement for cooperation that entailed specific presidential approval and tacit congressional approval.

technical infrastructure nor the materials for developing nuclear weapons, and would need several years to acquire both if it wished to do so. But what about the future, particularly if Israel's posturing on nuclear matters continued to arouse suspicion or if Israel actually declared a nuclear capability? Would not the expansion of nuclear energy programs in the Middle East only worsen an already inauspicious situation?

Concern has been expressed over the years that Israel has either developed nuclear devices or is far down the road toward a capability to develop them with a very short lead time.[13] Israel has substantial technical expertise in the nuclear field and is believed to have acquired enough weapon-grade plutonium from the unsafeguarded research reactor at Dimona (supplied by France under a 1957 agreement) for a small number of nuclear weapons. Israeli officials have denied that Israel has gone nuclear and have frequently stated that Israel will not be the first to introduce nuclear weapons in the Middle East.[14] But Israel has undertaken no specific international obligations, such as allowing IAEA safeguards at Dimona or ratifying the NPT, to help dispel concern.[15] Israel could, of course, put itself in a position of being able to produce nuclear weapons within a very short time after a decision to do so, if plutonium were available, without ever producing a complete device. The denial of nuclear weapon possession has not been extended to denial of a relatively short lead time capability.

Other questions arose in the wake of the presidential announcements in June. Would not opening this door to nuclear power reactor and fuel supply in the Middle East ease the way for the spread of nuclear technology and materials to less stable or radical regimes? Did not nuclear technology and materials ultimately add up to an

---

[13] See, for example, Pranger and Tahtinen, *Nuclear Threat in the Middle East*, pp. 11-20.

[14] See ibid., pp. 18-19.

[15] Most countries in the region, including Israel, have expressed support for the idea of a nuclear-free zone in the Middle East. Iran and Egypt have been the most active in this regard. Their approach suggests that the process should be advanced through consultations between the countries concerned and the UN secretary general. Israel's approach calls for direct negotiations with its neighbors on establishing a nuclear-free zone. Establishment of a meaningful free zone would no doubt entail safeguards on all nuclear activities in the region, including those at Dimona (a prospect which Israel has not publicly entertained to date). For positions on the basic free zone proposal, see Conference of the Committee on Disarmament, *Comprehensive Study of the Question of Nuclear-Weapon-Free Zones in All of Its Aspects*, Report of the Ad Hoc Group of Governmental Experts, CCD/467 (18 August 1975), pp. 20-21; and *Washington Post*, 1 October 1975, p. 4.

option to go nuclear, either by covertly diverting technology and materials from safeguarded peaceful activities to military programs or by overtly abrogating international obligations under agreements for cooperation? Was it necessarily only a matter of time, even if ten or more years, before several countries in this volatile area acquired, through peaceful programs, a capability to develop nuclear explosives? What about the possibility of an existing government being overthrown by a radical group? Last, but surely not least, what about the possibility of diversion, theft, or sabotage by terrorists?

All in all, the critics asked whether the United States should not seek to avoid or delay the introduction of nuclear energy in the Middle East. They questioned whether the energy situation of the states in the region justified moves toward nuclear power. Regardless of hopes for continued progress toward a peace settlement, the smoldering conflict, regional instability, and terrorist activity presented special security and political problems for any nuclear supply. For this and perhaps other reasons, congressional reactions to the presidential announcements contained a large dose of skepticism and some sharp criticism.

The case for responding to Egyptian and Israeli interest with a qualified affirmative—qualified in the sense that the United States would condition its offer on the acceptance by Egypt and Israel of additional protective measures—was presented by several spokesmen for the executive branch before two subcommittees of the House Foreign Affairs Committee.[16] The statements wove together political, economic, and arms control or security factors that favored United States entry into the proposed atomic energy agreements.

The presentations focused particularly on the chief concern in Congress—the security aspects or the risk side of the peaceful atom. From this standpoint, the executive branch stressed that the United States would provide for additional protective measures in its agreements with Egypt and Israel. Washington intended detailed provisions to ensure the following controls: (1) the location of facilities for fabricating and reprocessing nuclear fuel would be a matter of mutual agreement between the parties; (2) the location of plutonium stocks from spent reactor fuel would also be a matter of mutual agreement; and (3) mutual understandings would need to be reached on the

---

[16] See statements of the director of the Arms Control and Disarmament Agency, the director of the Department of State's Bureau of International Scientific and Technological Affairs, and the deputy assistant secretary of state for Near Eastern and South Asian affairs, in *Department of State Bulletin*, vol. 71, no. 1832 (5 August 1974), pp. 248-254; and statement by the under secretary of state for political affairs, in ibid., no. 1841 (7 October 1974), pp. 484-486.

application of adequate physical security measures against subnational or terrorist diversion, theft, or sabotage.

The third intention was obvious. The "matter of mutual agreement" in the first two was a diplomatic way of stating that Washington would have a definite and legal "say" on where spent fuel would be reprocessed and where recovered plutonium would be stored in the 1980s (when the proposed reactors would be completed and operating). Existing U.S. agreements for cooperation contain some controls on where fuel reprocessing may be performed and where nuclear materials, in excess of any required for use, may be stored. For the most part, however, these established controls seem merely to represent a basic requirement that safeguards be applied.[17]

The expressed conditions for supply to Egypt and Israel were designed to provide the United States with broader rights and greater scope for considerations other than the application of safeguards. For the foreseeable future at any rate, Egypt and Israel would probably have fuel fabrication and reprocessing services performed in the United States or Western Europe (although neither is currently in the commercial reprocessing business). The intended provisions in the agreements and accompanying diplomatic notes, plus statements by U.S. officials, made it clear from the outset that Washington certainly did not entertain the ideas of local reprocessing or local plutonium storage.

Added to this picture would be the usual controls regarding any transfer of U.S.-supplied materials or equipment to third parties, and the application of IAEA reporting and inspection procedures. If nuclear energy programs were to be introduced into the Middle East, this combination of safeguards and controls was considered the most that was necessary and negotiable by the administration. Keeping in mind that the proposed agreements would entail transfer only of reactors and low-enriched uranium fuel, the combined controls can definitely be seen as providing checks against potential security risks.

The crux of the case presented by the executive branch for the proposed agreements rested on still other grounds. Egypt, Israel, Iran, and perhaps other countries in the area had a not illegitimate interest in nuclear energy. There were other suppliers of nuclear reactors and fuel in the world, including France, the Soviet Union,

[17] For example, although there are variations in U.S. agreements for cooperation, a standard provision in many stipulates that "when any special nuclear material received from the United States . . . or any irradiated fuel elements containing fuel material received from the United States . . . are to be removed from a reactor and are to be altered in form or content, such reprocessing or alteration shall be performed in facilities acceptable to both Parties upon a joint determination that the provisions . . . [for safeguards] may be effectively applied."

and West Germany. (In late June, France agreed to sell several reactors to Iran, and, as it turned out, France, West Germany, and the Soviet Union were reportedly discussing reactor sales with Egypt later in 1975.) Washington could influence, to varying degrees, but could not control all the rules and players of the game.

Besides the important political implications of a negative response to the Egyptian and Israeli requests, and particularly the implications of campaigning against supply to the region in other supplier capitals, Washington's qualified affirmative rested on the assumption that regardless of the U.S. response, some suppliers would say "yes." Nuclear energy programs would be introduced in the region with or without the United States, and there was little reason to suppose that all potential suppliers would require as effective or extensive safeguards and controls as those proposed by Washington.[18] It was an old argument, not dissimilar to the one used for advanced conventional weapons transfers in parts of the world, but it still carried some weight.

Reservations about the proposed agreements in Congress did not disappear under the weight of the arguments favoring them. Not long thereafter, an amendment to the Atomic Energy Act was passed and signed into law that would, in effect, enable Congress to veto any proposed agreement for cooperation in the civil uses of atomic energy by majority votes on a concurrent resolution. Since such resolutions are not subject to presidential veto, this would make it easier than before to stop an agreement supported by the executive branch. In one sense, to be sure, the change was more cosmetic than real. Showdowns on proposed agreements had generally been avoided by close prior consultations between the executive branch (primarily the AEC) and the Joint Committee on Atomic Energy. Strong opposition from the committee normally meant the end of a proposal for the time being. But, with the issue of nuclear assistance and sales to Egypt and Israel on the front burner, this extension of congressional influence at least implied its exercise.

By mid-1975, Israel had reportedly lost its interest in pursuing nuclear reactor sales from the United States, but did not object to the proposed sales to Egypt. Israeli interest revived later in the year.

[18] With the example of France in mind, at that time, there was reason to expect that safeguards as extensive as those proposed by the United States would not be pursued by all suppliers. Past French exports had not required the application of IAEA safeguards. The general assumption in Washington, with no information from Paris to the contrary, was that French bilateral safeguards on their exports were not very thorough and stringent. As noted later, France's position has changed to one of applying international safeguards and other controls to future exports.

Progress toward an agreement with Egypt was made during President Sadat's visit to Washington in November 1975. Cairo continued discussions on nuclear reactor purchase with European suppliers, particularly France. Other actions in the area included: (1) progress in March 1975 toward the conclusion of a U.S.-Iran nuclear energy agreement for the supply of several large power reactors and associated fuel (which agreement may, as of late 1975, be encountering delay or difficulty, perhaps because of Iranian desires to acquire national reprocessing capabilities); and (2) Libya's arrangements with Moscow, a few months later, for assistance in developing a nuclear research center, along with Libya's apparently related ratification of the NPT.

A Broader Problem

In terms of its possible future derivation from technology developed or materials acquired through peaceful programs, the potential nuclear threat in the Middle and Near East regions was not diminishing. Nor was it diminishing in other parts of the world. If more explicit and additional protective measures are desirable in the proposed agreements with Egypt and Israel, the desirability of similar further controls in other sensitive areas becomes an unavoidable issue. Nuclear proliferation concerns are by no means limited to the special situation in the Middle East. Many considerations extending beyond that area went into Washington's reemphasis, in 1974 and thereafter, on the risk side of the peaceful atom and the need for more precautions.

The actions and reactions discussed in the two sections above reflect a few basic aspects of the nonproliferation problem. Nonproliferation considerations themselves, for example, can often pull in seemingly opposite directions on specific policy issues and must be thrown into the balance and weighed in light of broader political-military factors and interests. Each of the two cases highlights, although in a different way, a central concern that existing safeguards and controls provide something less than entirely adequate assurances against the risks accompanying the worldwide expansion of nuclear energy industries.

In mid-summer of 1974, Secretary Kissinger informed Congress that the administration was developing a comprehensive nonproliferation plan to be completed by the fall. One basic question—was nonproliferation still a viable policy stance?—was, in effect, instinctively answered in favor of reinforcing a longstanding U.S. interest in controlling the spread of nuclear explosive capabilities. This answer did not obviate the need for realistic assessments and contingency planning. But it did reflect the previously mentioned view that it is extremely difficult to see how world stability could be enhanced by the addition of more nuclear-weapon countries, while it is relatively easy to imagine how proliferation could increase regional (if not global) insecurity and instability.

How the United States proposed to deal with the "challenge" of realizing "the peaceful benefits of nuclear technology without contributing to the growth of nuclear weapons or to the number of states possessing them" was outlined by the secretary of state in his address before the United Nations General Assembly on 23 September. Washington's thinking on nonproliferation and related issues was presented in more detail by Senator Stuart Symington before the First Committee of the United Nations General Assembly the following month.[19] Taken together, the two statements reiterated much that was already established as U.S. policy, such as continuing, if not increased, support for the NPT, for international safeguards, and for the concept of regional nuclear-free zones.

Four new elements also appeared in the statements of U.S. policy: (1) high-level recognition of an urgent need to work towards the establishment of strengthened international safeguards and controls on nuclear technology and materials; (2) the way the United States proposed to strengthen existing safeguards and controls against potential diversion by national governments, particularly the note given to coordination and discussions with other nuclear supplier countries; (3) the U.S. proposal for an international agreement to enhance physical security measures against potential diversion or theft by individuals or nongovernmental groups such as terrorists; and (4) higher-level reiteration of the warning that the policy of promoting peaceful uses of nuclear energy was on the line if it meant proliferation of nuclear explosives.

The presentations, the sense of urgency, and the warnings may well have been, in large part, products of then recent events in South Asia and the Middle East. But the proposed policy outlines, which had been under consideration and review within the U.S. government for some time, represented neither a reaction to the Indian test nor a means of rationalizing nuclear assistance to Egypt and Israel. They stemmed, as noted, from a broader base. The steps suggested by the United States, along with other possible nonproliferation courses and principles, are the subjects of the remaining chapters.

[19] See Kissinger, "An Age of Interdependence," in *Department of State Bulletin*, vol. 71, no. 1842, pp. 501-502; and Release by Senator Stuart Symington's Office, "U.S. Statement on Nuclear Issues," 21 October 1974, pp. 1-10.

3
APPROACHES TO REDUCE SECURITY RISKS

What can be done to reduce proliferation and other security risks associated with nuclear energy programs? Answers to this broad question necessarily come in several parts. The most important include controls by and cooperation among nuclear supplier countries and cooperation between suppliers and users. The security risks of the peaceful atom can be reduced, in other words, both (1) by the establishment of further restraints on the commercial-political competition among supplier countries themselves, or the acceptance of additional basic parameters to the supply game, and (2) by the promotion of positive alternatives, particularly multinational endeavors, to the trend toward independent national programs and developments. Although these two policy packages are discussed separately for the purposes of analysis, they are obviously interrelated in many respects.

The issue of reaching, or attempting to reach, some understandings or agreements among present and potential suppliers of nuclear materials, equipment, and technology was first on Secretary Kissinger's program announced in September 1974 for strengthening nuclear safeguards.[1] One month later, in New Delhi, he noted that India has an "important role" in the effort to reach multilateral agreement on restraints on nuclear exports.[2] On Martinique, in December, President Ford and President Giscard d'Estaing "explored how, as exporters of nuclear materials and technology, their two countries could coordi-

[1] See Kissinger, "An Age of Interdependence," in *Department of State Bulletin*, vol. 71, no. 1842, p. 501. Major nuclear supplier countries, besides the United States, include Canada, France, West Germany, Great Britain, and the Soviet Union. Other suppliers (now or in the near future) include Belgium, India, Italy, Japan, the Netherlands, South Africa, Spain, and Sweden.

[2] Henry A. Kissinger, "Toward a Global Community: The Common Cause of India and America," address before the Indian Council on World Affairs (28 October 1974), in ibid., vol. 71, no. 1848 (25 November 1974), p. 743.

nate their efforts to assure improved safeguards of nuclear materials. This exploration alone represented a significant threshold.[3]

The fact that discussions have been taking place among the key supplier countries on general restraints and specific issues is not secret. Although the details of the talks have remained so to date, the outlines of agreement among the major suppliers on principles to govern nuclear exports have, as noted, recently been announced. These principles relate to, among other things, the application of international safeguards, further restraint in exporting sensitive technologies, and commitments to strengthen requirements for protecting exported materials. Still other areas and issues, under the heading of supplier controls and understandings, need to be considered.

Supplier Controls and Understandings

Countries which desire and can afford independent nuclear fuel facilities will likely continue in that direction. Tendencies toward national facilities will exist even in countries where independent nuclear fuel cycle programs are scarcely, if at all, justifiable in economic and energy terms. Regardless of the degree of success (or the lack of success) in channeling such impulses into multinational undertakings, there remain several areas where the proliferation and other security risks of the peaceful atom could be reduced through increased nuclear supplier understandings on controls and coordination.

International Safeguards. A basic understanding should continue to be sought among present and potential supplier countries that have not already agreed to make all exports of nuclear materials, equipment, and technology subject to the application of IAEA safeguards. Exports of types and quantities for medical or other purposes, which by common agreement do not require safeguards, would be excepted. (How to sustain and further strengthen the current IAEA safeguards system is a subject of the following chapter.)

Unanimous agreement among nuclear suppliers on the application of IAEA safeguards can best ensure that no competitive advantage in the nuclear market will be sought through a weakening of

[3] "Text of Communiqué (16 December 1974)," in ibid., vol. 72, no. 1855 (13 January 1975), p. 43. The significance of this statement stems from the view that it appears to indicate a more cooperative French approach than in the years past, or less of a tendency on the part of France to stand off or go its own way, in nuclear export and safeguards matters. France has become more forthcoming on the issues of applying IAEA safeguards and other controls to future exports. What its and other suppliers' positions will be on still outstanding issues remains to be seen.

safeguards requirements. IAEA safeguards, as noted in the following chapter, have definite limitations that need to be recognized. But they are much better than purely paper guarantees, which have no place in the safeguards business, and certainly better than no safeguards at all.

Article III of the NPT, as noted, has resolved a major portion of the safeguards problem. It requires not only that non-nuclear-weapon states party to the treaty accept IAEA safeguards on all their peaceful nuclear activities, but also that no party will provide "source or special fissionable material," or related processing and production equipment, to any non-nuclear-weapon country unless IAEA safeguards are applied to that material and equipment. The Zangger Committee (a group of supplier countries cooperating through the IAEA) has, in addition, worked long and hard on developing and gaining acceptance for a "trigger list" of export items that would require the application of safeguards.

Most of the present and potential nuclear exporters are either now party to the NPT or expected to become party to it in the near future. Problems can arise, however, because some present (France) and potential (Spain, South Africa, and India) suppliers are not party to the treaty and are not expected to join in the foreseeable future, and because several present and potential importers of nuclear materials and equipment (for example, Argentina, Brazil, Egypt, India, Israel, Pakistan, and South Africa) are not party to the treaty. Problems also arise because there are gaps in the technical approach of the Zangger Committee and the "trigger list."

Parts of these problems were resolved in late 1974 when thirteen countries, including most of the major nuclear suppliers today, endorsed the "trigger list" and related understandings.[4] This added up to agreement to condition the export of "trigger list" items, as a minimum, on a recipient's acceptance of IAEA safeguards on the imported items, promise not to use the supplied material for any nuclear explosive purposes, and "assurances that the imported items will not be re-exported in a way that would circumvent NPT objec-

[4] Australia, Canada, Denmark, Finland, West Germany, Great Britain, the Netherlands, Norway, the Soviet Union, and the United States exchanged letters and sent notes to the IAEA director general informing him of their agreement in August. East Germany, Hungary, and Poland sent notes the following month. The United States forwarded two supplemental letters, one of which set forth U.S. policy to require safeguards on items in addition to those on the "trigger list." See International Atomic Energy Agency, INFCIRC/209, 3 September 1974.

A few other countries—including Austria, Japan, Sweden, and Switzerland—are generally expected to endorse the "trigger list" and related understandings. Belgium has publicly stated its intention to do so.

tives."[5] These actions narrowed potentially wide gaps in the IAEA safeguards system. Over a year after this development in the Zangger Committee's efforts, a significant event was reported in the context of the multilateral suppliers' discussions initiated by the United States—French agreement to require, as a general rule, IAEA safeguards in conjunction with its nuclear exports.

Despite these favorable safeguards developments, a paradox still stands out in the nuclear supply business. Non-nuclear-weapon states party to the NPT have assumed an obligation to accept safeguards on all their peaceful nuclear activities, whether indigenously developed or based on imports. Nonparties to the treaty are under no such obligation, and some (witness Brazil and India) represent importers of substantial amounts of nuclear materials and equipment. Imported materials and equipment subject to safeguards cannot legally be used in unsafeguarded indigenous programs, but the technical infrastructure or technicians trained through programs based on imports can.

What to do about this paradox remains unresolved. One frequent suggestion has been that nuclear suppliers should condition their exports to countries not party to the NPT on a requirement that they accept safeguards on all their nuclear activities. At present, a consensus on applying this suggested export condition could not likely be reached among all nuclear supplier countries and, in the absence of agreement, it would be difficult for a supplier country to apply such a condition consistently. Nevertheless, some major suppliers (specifically Canada, Great Britain, and the Soviet Union) might well agree to its application on their exports, perhaps particularly if the United States were to give it active support. On balance, the proposed safeguards condition or some variation thereof seems to merit further discussion among supplier countries, as a desirable objective, so long as it does not detract from the possibility of reaching more meaningful technology controls.

Physical Security Measures. Agreement should also be sought among nuclear supplier countries that have not already committed themselves to an export requirement that any country receiving nuclear materials, equipment, or technology have adequate physical security measures in effect. The enhancement of physical security measures—the guard against the risk of nongovernmental nuclear diversion, theft, or sabotage—is a subject of the following chapter.

[5] William O. Doub and Joseph M. Dukert, "Making Nuclear Energy Safe and Secure," *Foreign Affairs*, vol. 53 (July 1975), p. 760.

Special Precautions. The security risks of the spread of the peaceful atom could also be reduced by understandings among supplier countries that nuclear exports to certain areas or countries, especially where there is serious regional conflict or national instability, should either (1) require special precautions besides existing international safeguards and adequate physical security measures or (2) be avoided.

The cases of Egypt and Israel come readily to mind, as the Middle East is an area where Washington adopted additional protective measures as conditions for nuclear supply from the United States and into which other nuclear suppliers are moving. But this area and these countries are, as noted, by no means in an entirely unique situation when it comes to the desirability of extra precautions. Like it or not, different countries present a different complex of security risks relative to nuclear exports, and especially to such matters as (1) the transfer of uranium enrichment or reprocessing technology, equipment, and facilities, (2) the location of plutonium stocks from spent reactor fuel, (3) the transfer of high-enriched uranium and HTGRs, and where related fuel fabrication services are to be performed, and (4) the transfer of plutonium and FBRs, when they come on line, and where plutonium fuel fabrication services are to be performed for FBRs or for plutonium recycle in LWRs.

Export, for example, to countries such as Belgium, West Germany, Italy, Japan, the Netherlands, Sweden, and Switzerland is one thing. These non-nuclear-weapon countries either adhere to the NPT or have a positive attitude toward adherence. More important, under existing and foreseeable circumstances, they have little if any real incentive to acquire nuclear explosive capabilities—not to mention the substantial disincentives to their doing so.

Equivalent security perspectives, disincentives, and attitudes do not apply to all countries. What about such nonparties to the NPT as Israel (which is believed to have the technical capability and nuclear materials for a few nuclear devices already, and whose security concerns need not be recounted), Egypt (which has signed the NPT, but made ratification dependent upon Israel's adherence, and which naturally watches Israel's every move), Pakistan (which occupies an uneasy position next door to India's nuclear developments), and Argentina and Brazil (both of which have expressed an interest in developing nuclear explosives for peaceful purposes)? Even some present parties to the NPT—for example, Iran, South Korea, or Taiwan—may be cause for concern somewhere down the road depending upon their security perspectives.

Since most countries still rely on imports for their nuclear power programs, nuclear exporters can, with cooperative understandings on special precautions, exercise some control over the security risks of a spreading peaceful atom. Such measures could provide assurances not only to the exporting countries but also to countries in the region of the recipient. Supplier agreement on additional precautions in certain areas, including export denial in special cases, need not be viewed as simply negative. The promotion of regional multinational facilities, as discussed in the following section, would represent a positive alternative for satisfying nuclear fuel cycle needs.

Nevertheless, broad agreement on special precautions will be very difficult to achieve because of the varying political, economic, and military interests of the nuclear exporting countries, and because of differing government-industry relations. Many, if not most, questions on export plans or proposals of other suppliers will consequently continue to be discussed, debated, and dealt with on a case-by-case basis. This will make progress in the following area all the more critical.

General Supplier Cooperation and Coordination. There is a basic need for better coordination of plans, more extensive consultations, and more detailed information exchange between nuclear supplier countries.

Coordination, consultation, and information exchange on bilateral and multilateral bases have improved substantially in the past year. But greater efforts in this direction could continue to (1) reduce the possibility and probability of significant misunderstanding among suppliers, even under widely accepted competitive ground rules, (2) allow for an airing of differences of opinion or policy before issues come to a head, or before questionable export proposals become, in effect, irreversible commitments, and (3) facilitate the resolution of policy differences on problems such as additional protective measures or transfers of sensitive technologies to certain areas, or at least minimize the adverse impact of policy disagreements. Finally, increased coordination and rationalization of plans among supplier countries on how the burgeoning market for uranium enrichment services will be met could avoid unnecessary duplication and abet multinational approaches.

In several of these respects, the practices of the United States, as well as of others, have clearly left something to be desired. Given Washington's view that additional precautions should be applied to exports to the Middle East, would it not, for example, have been much better (and possibly productive) to have consulted with other nuclear supplier countries before opening the export door to the area through a presidential announcement? In Washington, notice of a third country's search for a reprocessing capability from other nuclear suppliers has, all too often, arrived first through indirect channels or intelligence sources.

Both the viewing of nuclear energy as "political chips" and the competing commercial concerns have frequently stood in the way of—and must be largely overcome to achieve—effective supplier consultation, cooperation, and coordination. These are three very overworked "c's." But their exercise and any useful results could add up to a valuable curb on the security risks of the peaceful atom, especially in the absence of other agreements suggested herein.

NPT Carrots. A few other subjects of secondary importance (measured against the primary importance of the preceding four topics and the following section on multinational approaches) may be worth pursuing at some stage in supplier country discussions or through regular diplomatic channels. One such suggestion would be for suppliers party to the NPT to develop, where possible, additional benefits accompanying adherence to the treaty.

If most NPT parties have not really given up much—if anything—by joining, neither have they received much—if anything—in the way of nuclear assistance that they would not have received at any rate. The United States has since late 1974 taken a few steps in the direction of preferential treatment for NPT parties in the areas of IAEA medical research and technical assistance programs. The possibility of more substantial benefits, such as some kind of preferred access to uranium enrichment services, deserves further consideration.

Three important factors, however, need to be taken into account in examining possible NPT "carrots" or benefits accompanying adherence. First, while a country's attitude toward the NPT is an important aspect in the examination of nuclear export proposals, any preferential treatment for parties to the treaty should not discount the necessity of applying other central criteria. Second, no practicable "carrots" will likely alter the attitudes of nonparties on the question of ratification. Third, special restrictions against countries, only on grounds that they are not party to the treaty, appear undesirable generally and probably unacceptable to most nuclear supplier nations. Nevertheless, making treaty membership more attractive is another matter and could make for a healthier NPT regime.

Problems and Prospects

Is there sufficient incentive, recognition of common interests and common dangers, and political will among present and potential nuclear supplier countries to bring about some effective basic understandings, formal or informal, on further controls and coordination? Several difficulties and obstacles which need to be overcome if this question is to be answered in the affirmative have been described in the preceding pages. Each proposal will undoubtedly present particular problems, as well as benefits, to each supplier country. Achievement of additional understandings regarding the nuclear supply business will require, in every supplier capital, determined political leadership over the technical discussions and private interests.

The attitudes of the various political leaderships toward further controls and outstanding issues, as well as the interplay between political and private forces in different countries, remain largely to be seen. But, besides evident obstacles, there are some hopeful signs in addition to the recent agreement. Of the principal foreign suppliers today, Great Britain and Canada are strong supporters of nonproliferation generally and have indicated an active interest in further nonproliferation measures, such as the control of exports of sensitive technologies to certain areas. The Soviet Union has supported and participated in previous nonproliferation efforts, and there are no signs that this interest has diminished as the Soviet Union has become more active in the international nuclear market. Its nuclear research center arrangements with Libya interestingly coincided with Libya's ratification of the NPT.

Understandings with West Germany and especially with France will likely encounter much more difficulty. Despite serious reservations in Washington, Bonn concluded a complete nuclear fuel cycle agreement with Brazil, and Paris has appeared willing to sell reprocessing capabilities to Pakistan and South Korea. (South Korea just recently confirmed that it has canceled transactions to purchase a reprocessing facility.) On the other hand, West Germany has been cooperative on other nonproliferation issues and, perhaps owing in part to Washington's expressed concerns, built several controls into its agreement with Brazil.[6] France apparently raised no serious objec-

[6] The controls in the West Germany-Brazil agreement are similar to those applied by the United States on its exports of nuclear materials and equipment. See "German Stand on Nuclear Accord," letter to the editor from Dr. Niels Hansen, Chargé d'Affaires, Embassy of the Federal Republic of Germany in Washington, *New York Times*, 7 July 1975, p. 24. The application of these controls does not, of course, erase the fundamental difference regarding the export of reprocessing and enrichment technologies or equipment in the first place.

tion to efforts to channel South Korea's interest away from acquiring a national reprocessing capability. This, along with France's previously cited willingness to explore coordination of efforts to improve safeguards and its willingness generally to apply IAEA safeguards to its exports, are notable indications that France under the presidency of Valéry Giscard d'Estaing has become and may well continue to be considerably more cooperative than in the past.

As for the United States, Washington has focused high-level political attention on the need for further controls, taken substantial steps toward putting its own house in better order, and at least partially accepted some of the implications of its faded monopoly in nuclear energy affairs. But more needs to be done. Continued high-level political attention and intermittent intervention at this level will be necessary both to influence other suppliers and to keep Washington's own programs on track. The influence of the United States in nuclear matters will depend largely upon Washington's conduct and example and upon its understanding of the real implications of a fading monopoly. Among other things, these implications call for a cooperative attitude toward the application of IAEA safeguards on U.S. civil nuclear activities, for increased cooperation with allies in nuclear energy research and development, for a continuing willingness to assist in allied efforts to gain some independence in nuclear fuel supply, and for an understanding of the longstanding suspicions in western Europe that the overriding concern of the United States is to protect its own competitive position by "made in U.S.A." solutions or other means.

The present and prospective nuclear market, worldwide, runs into many billions of dollars. Given the commercial and balance-of-payments benefits in the export of power reactors, enrichment services, and other nuclear materials, equipment, or technology, there is and will remain a natural tendency among nuclear supplier countries to gain as large a portion of the export market as possible. This tendency and the strong inclination on the part of exporters and importers to view, subsidize, and otherwise support nuclear energy as "political chips" need to be curbed if the spread of the preconditions for nuclear explosive capabilities is to be restrained, and the security risks of the peaceful atom reduced.

Washington's record on curbing these tendencies is by no measure completely clean, but neither are the records of other nuclear supplier countries without some smudges. Supplier restraints and understandings will, by definition, mean give-and-take on all sides. Just as Washington cannot expect to use any supplier understandings as a means of helping buttress its present (but passing) preeminence in the peaceful applications of nuclear energy, so other suppliers cannot expect Washington to bargain away its current competitive edge. Nor can Washington be expected to refrain from questioning commercial practices that raise serious proliferation issues.

Although this section has concentrated on the problems and prospects of nuclear suppliers' reaching further understandings among themselves, another problematical aspect has to be considered: Will certain understandings among suppliers serve to increase the incentive for indigenous developments in third countries? Indigenous development of a chemical reprocessing capability, for example, would in almost all cases cost more and take longer than outside supply of that capability. At the same time, unsafeguarded indigenous developments (either overtly by a nonparty to the NPT or covertly by a party) could introduce more problems than would have occurred under export from a supplier country.

This potential disadvantage may or may not turn out to be the case, depending upon the circumstances for each country. In general terms, implementation of the suggested supplier understandings would not hinder the reasonable spread of the benefits of nuclear energy to developing and developed nonsupplier countries. Such agreements would entail no unnecessary economic costs, while the multinational approaches proposed in the following section would in fact achieve economies of scale.

It is fair to assume, therefore, that the supplier understandings in practice would neither significantly increase impulses toward indigenous programs across the board nor result in any particular situation that would have been less risky had development been facilitated by outside supply. Moreover, in situations especially risky from a proliferation standpoint, a supplier's ability to control all uses and spin-offs of his exports through today's standard safeguards, guarantees, and nonreplication provisions should not be overestimated. Nor should the value of the delay imposed by indigenous nuclear development be underestimated.

The idea of understandings on controls and coordination among nuclear supplier countries is not new. A 1970 RAND report presented a cogent case "to restore a balance between safeguards and technology transfer control." Although parts of the authors' analysis have since been overtaken by events, much remains valid today:

> There is little that can be done directly to slow the development of nuclear technology in the advanced countries, although it can perhaps be channeled to some extent through

> cooperative U.S. actions. We should . . . like to see civilian nuclear programs closely interdependent and based in critical areas on multinational facilities so as to reduce opportunities for transfer of civilian nuclear materials and facilities to military purposes.
>
> To effect significant control over the distribution of potentially dangerous nuclear technology to the less developed countries, the U.S. needs the help of the . . . important suppliers of nuclear technology.[7]

Nonproliferation efforts directed at what the authors term "inhibiting the preconditions for rapid development of nuclear weapons programs" are still necessary and have grown more urgent today. They should constitute a major portion of any program aimed at reducing the proliferation risks of expanding nuclear energy industries.

Multinationalism: A Regional Route

The emerging nuclear energy picture, as noted, includes sharply rising demands for uranium enrichment services in the years ahead, and expanding chemical reprocessing capabilities. In several advanced industrial countries with sizable nuclear energy programs, national facilities for performing these services (especially enrichment) will appear attractive and logical on political, prestige, technical, nuclear energy independence, and perhaps some economic grounds. Despite these attractions for advanced nuclear energy consumers, independent national facilities are not necessarily the desirable alternative from the economic standpoint (depending, for example, on scale) and the standpoints of nonproliferation, security, and safeguards objectives.

Development of national capabilities becomes highly questionable when such enterprises appear contrary to logic and economics as they relate to a country's level of nuclear energy programs. Pakistan, South Korea, and Taiwan, for example, have expressed interest in acquiring reprocessing capabilities even though their present and prospective nuclear energy capacities are relatively small. While political, prestige, and technical considerations alone can be driving forces in defining interests for national reprocessing or enrichment facilities, such pursuits naturally raise, or compound existing, ambiguity about intentions regarding potential military applications or options.

[7] V. Gilinsky and W. Hoehn, *Nonproliferation Treaty Safeguards and the Spread of Nuclear Technology*, RAND, R-501 (May 1970), pp. vi-vii. See also, pp. 24-36 for further analysis on these points.

The spread of national enrichment or reprocessing capabilities, regardless of size, presents problems for nonproliferation and safeguards objectives. The question how to control this spread and reduce the problems can be answered, at least in part, with commercial-size, multinational regional facilities or, preferably, regional nuclear fuel cycle "parks." These would entail the co-location of multinationally owned and managed enrichment, fabrication (especially any plutonium or high-enriched uranium fabrication), and reprocessing facilities, combined with multinationally controlled plutonium stockpiles.[8] The NPT Review Conference in May 1975 gave attention not only to the importance of safeguards but also to the potential promise of multinational nuclear enterprises. The economic, security, and other advantages of such a course, which takes into account the growing worldwide interest in nuclear energy, are substantial.

First, the application of international safeguards to a few multinational regional facilities would present far less difficulty than their application to numerous national facilities, and would involve lower costs (a not inconsiderable factor given the prospective sharp increases in costs necessary to support safeguards functions).

Second, facilities which were multinationally managed and operated, in contrast to facilities which simply had financial backing from outside countries, would have built-in checks and balances against either diversion or takeover by any single country. (Plutonium, for example, would remain in multinationally controlled storage banks until required for use in energy programs.)

Third, multinational regional facilities would achieve economies of scale, would be more cost-effective than national facilities in developing areas where no single national nuclear energy program is likely to provide economic justification for the acquisition of either enrichment or reprocessing capabilities, and would incur no substantial additional costs in developed areas (despite the increased transportation requirements in both cases).[9]

---

[8] Subsequent to this writing, Secretary of State Kissinger announced a United States proposal for "the establishment of multinational regional nuclear fuel cycle centers . . . [to] serve energy needs on a commercially sound basis and encourage regional energy cooperation." See Henry A. Kissinger, "Building International Order," address before the United Nations General Assembly (22 September 1975), in *Department of State Bulletin*, vol. 73, no. 1894 (13 October 1975), p. 551.

[9] This advantage is especially valid for enrichment plants based on the gaseous diffusion process because of their large capital and electricity requirements. The argument may admittedly be altered by the advent of commercially competitive and smaller-scale centrifuge plants. But even if centrifuge or other techniques

Fourth, the concept could be put into practice on a broad basis, including (perhaps especially in the Pacific region) practice by the more advanced nuclear countries, while it avoids (through participation of the nuclear "have-nots") much of the inherently discriminatory nature of most nonproliferation measures.

Finally, threats of subnational diversion, theft, and sabotage would be minimized, given adequate physical security and materials accounting measures at co-located facilities. Transport outside the "park" would consist of the incoming, heavily shielded casks (running into several tons) of highly radioactive and unusable (until reprocessed) spent fuel, and the outgoing fully fabricated fuel rods.

These advantages add up to a conclusion that a key part of the discussions among nuclear exporters should focus on reaching agreement to encourage multinational enterprises and perhaps to limit supply of any chemical reprocessing or uranium enrichment technology and equipment to them. Active promotion of this concept, and actual practice by the promoters, may well be required if regional nuclear fuel facilities with multinational participation are to be realized on any broad scale.

Although this section has focused on the main aspects of the nuclear fuel cycle in general terms, both international promotion of multinationalism and relevant internal decisions should avoid placing a reprocessing-cart ahead of an enrichment-horse. The necessity and desirability of reprocessing in the near future, or before the lead time relative to fueling breeder reactors with plutonium, have been the subject of considerable debate. Reprocessing of spent fuel for plutonium recycle in current reactor types can, as noted, fulfill about 10 to 20 percent of the need for nuclear fuel by the 1990s. At the same time, the plutonium product means additional security precautions, costs, and risks.

The interest of other countries in reprocessing cannot, of course, simply be ignored. But, for the next several years at any rate, the objective importance of reprocessing will depend largely on the availability, security of supply, and costs of uranium and uranium enrichment services. In the next fifteen years much more enrichment capacity, or many more plants, will be necessary to meet world demands. With these considerations in mind, along with the outlook

become commercially competitive on smaller scales, they would still be fairly large (and certainly competitive on larger scales) and complicated to operate. Thus, the development of such advanced techniques by no means automatically invalidates the stated advantage of the cost-effectiveness of multinational undertakings. Moreover, it should be kept in mind that uranium enrichment services constitute only a very small percentage of the costs of nuclear energy.

for relatively small centrifuge facilities, the promotion of multinationalism will need to focus increasingly on enrichment or the front end of the fuel cycle.

More about Problems and Prospects

The present picture indicates that there may yet be an opportunity to control the spread of independent national enrichment and reprocessing capabilities if agreement can be reached among suppliers actively to encourage multinational enterprises, and, preferably, to make "multinationalization" a precondition for supplying such technologies and equipment. The only countries today with operational or near-operational uranium enrichment facilities (pilot plant or commercial scale) are the United States, the Soviet Union, Great Britain, France, the People's Republic of China, West Germany, the Netherlands, Japan, and South Africa. Reprocessing capabilities, however, are not as limited. Countries with commercial or pilot-plant-scale reprocessing facilities in operation or under construction include all the above, except the Netherlands and South Africa, plus Argentina, Belgium,[10] Italy, and Spain. Canada and Sweden have plans to construct commercial reprocessing plants, and Brazil has made arrangements to acquire reprocessing capabilities from West Germany.

The concept of multinational regional facilities represents, in some respects, a halfway house between the "internationalization" foundations envisaged by the Baruch Plan and what developed in its place—the more convenient and politically acceptable unilateral and bilateral programs. Between these two poles, a multinational impulse has persisted in thought, proposals, and, to a degree, in practice.

But if the multinational concept is not new in the nuclear field, its active advocacy by the United States is. It can be argued, with fair justification, that Washington should have pushed "multinationaliza-

[10] The small reprocessing pilot plant in Belgium, apparently being phased out of operation, is a cooperative facility with participation by the several western European countries (including France and Germany) of the Nuclear Energy Agency of the Organization for Economic Cooperation and Development. As a commercial company, with members participating as shareholders and the governments represented in the supervisory process, it may be considered (as suggested by Henry Nau in a letter to the author) as a kind of prototype structure for cooperation in nuclear fuel areas. It is also worth noting, in this regard, that Great Britain, West Germany, and France agreed in 1971 to coordinate their reprocessing plans with the aim of having the availability of reprocessing services correspond to actual demands.

tion" or reintroduced "internationalization"—divorced from the question of nuclear superpower arms control[11]—years earlier in the 1960s. At that time, near monopoly and clearer preeminence in commercial nuclear matters could have provided the United States with greater bargaining leverage than it has today. What cannot be disputed is that this advocacy has come none too soon if there is to be any chance of channeling nuclear fuel facilities, either in whole or in part, into regional multinational undertakings.

Despite the advantages of the multinational approach for commercial-size facilities, as well as the still limited spread of the relevant technologies and the active advocacy of the United States, potential disadvantages and considerable difficulties stand between the idea of "multinationalization" and its being realized on any significant scale. Even when such an objective is accepted in principle by all the prospective participants in an enterprise, problems such as how to share production benefits and management and operations responsibilities offer no easy solution. Different sets of interests among those entertaining a multilateral undertaking and widely varying industrial and industry-government relations within countries are only two of the basic ingredients which can complicate, if not prevent, attempted cooperation.[12]

Technology transfer and possibly waste disposal issues represent two other sizable problems in multinational approaches. It may be

[11] For a recent proposal advocating "internationalization" of reprocessing facilities, plutonium and high-enriched uranium, uranium enrichment facilities, and PNEs, see Lincoln P. Bloomfield, "Nuclear Spread and World Order," *Foreign Affairs*, vol. 53 (July 1975), pp. 743-755. He correctly concludes that "[f]or all parties, even the partial approach [or step-by-step approach toward "internationalization"] calls for a leap of imagination. But so does living in a world of proliferated nuclear weapons where every minor international quarrel could become genocidal."

"Internationalization" has not been considered further herein for several reasons. Realization of the regional multinational approach on a relatively broad scale would have much the same advantages without encountering the same degree of difficulty (although the degree of difficulty would still be very considerable). Compared to "internationalization," the multinational approach does not run as counter to nationalistic tendencies generally. Also, "internationalization" would arouse significant opposition, whether or not well founded, from industrial and commercial concerns in several major countries with advanced nuclear programs, whereas "multinationalization" may actually be supported by many such concerns. Finally, "multinationalization" does not necessarily preclude future steps toward some form of "internationalization."

[12] For an analysis of how different perceptions of government-industry relations, and different industrial traditions generally, can contribute in large part to the failure of attempted cooperative projects in the nuclear field, see Henry R. Nau, *National Politics and International Technology: Nuclear Reactor Development in Western Europe* (Baltimore: Johns Hopkins University Press, 1974), pp. 81-90 and 212-235.

difficult to achieve the cooperative participation of several countries in multinational undertakings under rules of limited or no access to sensitive technologies. Yet full access would eventually add up to transferring a capability to construct national facilities. Precautions necessary to protect the sensitive technologies of the nuclear "haves" may well prove to be stumbling blocks to "multinationalization." Nuclear waste might likewise prove a stumbling block for any commercial-size multinational reprocessing facilities since these could, in effect, involve asking one country to store the long-lasting and highly radioactive waste from other countries' nuclear reactor programs.

Far more substantial obstacles to realizing the regional multinational approach—in addition to those general difficulties discussed earlier in the chapter—exist in the tendencies toward independent national programs and plants in advanced nuclear supplier countries and in nuclear importing countries. These tendencies, as noted, may be based on such considerations as economic nationalism, political prestige, technological accomplishment, and energy independence. Military-security considerations may well come into play, however quietly, in some countries. Nationalist tendencies can also be etched deeply in a pattern of bureaucratic inertia in governments and industries.

While economic efficiency itself argues for interdependence and some form of multinational participation in major fuel facility undertakings, there are two respects in which concerns about comparative competitive postures—what may be called political-economic nationalism—can work toward national facilities. These political-economic forces represent a significant impediment (1) to multinational enterprises among competing nuclear supplier countries which can afford independent programs (witness the differences between France and Germany on uranium enrichment matters), and (2) to nuclear supplier cooperation in encouraging multinational approaches in other parts of the world and gearing export policies to this end (witness the recent West German agreement to provide the entire nuclear fuel cycle to Brazil).

Economies-of-scale arguments will thus encounter competing considerations even in cases of commercial-size facilities, where economic efficiency clearly favors multinational approaches. Moreover, these arguments do not apply to anywhere near the same degree for possible small-scale, pilot-plant operations as they do for commercial-size facilities. The initial interests of Pakistan, South Korea, and Taiwan in reprocessing have centered on the former kind of facility. The prospect of the spread of relatively small enrichment projects, such as centrifuge plants, also looms on the horizon of nonproliferation and safeguards problems.

The spread of small-scale facilities is one of the more immediate policy problems. The most appropriate response to it would be agreement among nuclear suppliers to promote multinationalism and to avoid providing reprocessing or enrichment technology and equipment in areas where these capabilities have little or no justification in terms of a country's nuclear energy programs. (At a minimum, in cases where agreement to refuse supply is not in the offing, reprocessing or enrichment assistance to third countries should be conditioned on supplier involvement in the operations and management of the facilities. A binational approach would entail no significant if any extra costs, and would have substantial advantages from the standpoint of nonproliferation and safeguards objectives.)

A main challenge of the multinational approach is whether a multiplication of relatively small national programs can be avoided, at least in part, by giving many countries a "piece of the action" in larger undertakings. The outcome will depend not only on whether potential importers or developers can be satisfied with only a part of the whole, but also on the ability of nuclear suppliers to reach understandings on promoting multinationalism and limiting or refusing assistance for single nation projects.

Testimony to the existence of hurdles to be overcome, if the multinational approach is to take hold, needs to be read in the light of other developments. First, a form of multinational participation, specifically in financing, is working its way in American and French uranium enrichment plans. The development of the centrifuge by West Germany, Great Britain, and the Netherlands is tied together in a multinational structure. The United States also has a sharing proposal for multinational enrichment projects on the table.[13] Second, although "multinationalization" of existing and prospective facilities in the advanced nuclear nations will remain desirable in setting an example, it is not the crux of the problem for proliferation, security, and safeguards. Finally, stumbling blocks or impediments to multinational approaches may well be less operative in some parts of the world, such as the Pacific region, than in others.

[13] In 1971, the United States proposed (although in a seemingly cautious manner) sharing its most advanced diffusion technology with allies for the construction of a multinational uranium enrichment plant or plants abroad, subject to arrangements for the application of international safeguards and technology controls. This offer was apparently expanded, at the Washington Energy Conference of February 1974, to include other more advanced technologies such as the centrifuge.

To a certain degree, "multinationalization" represents a structural approach to the essentially nonstructural problem of nuclear proliferation. The proposed structures, nonetheless, have several significant advantages in their own right. Even partial success in the pursuit of regional multinational facilities could serve to reduce the security risks and safeguards costs of the peaceful atom. Without active encouragement by—and understandings to restrain a competitive-commercial urge among—nuclear suppliers, the chances of regional multinationalism taking broad hold will be slim. The roads to specific multinational enterprises and to related supplier understandings will not be easy under foreseeable circumstances. This fact highlights the importance of supplier understandings, discussed earlier, on controls and coordination.

4
FURTHER PROPOSALS AND POLICY ISSUES

Enhancing Measures against Subnational Threats

The vulnerability of modern society to all kinds of extremely nasty threats needs no recounting. In addition to problems associated with nuclear materials, highly toxic chemicals abound in the industrial world, and varieties of lethal biological agents are commonplace in many laboratories. The safeguards against subnational chemical or biological threats becoming reality are, in several respects, much less developed than the relative high degree of precautions currently applied to the nuclear industry. While considerable attention has been focused on the subnational nuclear threat in recent years, this may not be the worst, or the most likely, threat to occur. It demands, nevertheless, continuing attention and increased protective efforts.

The construction of a workable atomic bomb is not beyond the capability of a determined and technically competent individual or group with a few kilograms of plutonium or high-enriched uranium. Very small quantities of plutonium, much less than one kilogram, could be attached to a few pounds of standard high explosives to create a grave public health-contamination threat. The forecast growth of the nuclear power industry adds up to significantly increased amounts of high-enriched uranium and to vast quantities of plutonium in transit and storage throughout the United States and abroad. The scope and nature of terrorist activities in modern times are expected to continue, if not increase. So are radicalism and irrationality, individual or group.

The potential sums of such basic considerations, however unpleasant and unthinkable, are by no means impossible. They have produced, particularly in the United States, heightened concerns about and keener appreciation of the threats of subnational nuclear diversion, theft, or sabotage. With doubts growing outside and inside the government about the adequacy of controls then existing against unauthorized possession or use of nuclear materials within the United States, a governmental review of domestic safeguards was undertaken at presidential direction in 1970.

The study concluded, in effect, that the general protection measures required by the AEC for the licensed or private sector no longer, if ever, provided sufficient assurance against potential threats becoming reality. What seemed to suffice for a nuclear industry through its infancy and childhood stages was patently not enough for a full-grown enterprise. Several actions, on the home and export fronts, resulted from this and other reviews in Washington.

First, the AEC recommended and promulgated, in late 1973, new and more specific regulations which substantially upgraded the protection of nuclear materials in the U.S. private sector. This upgrading included more frequent and improved methods for materials accounting procedures (designed to detect covert diversion of any significant quantities) and—of major importance—more explicit and extensive physical security measures (designed to prevent overt or covert theft by immediately detecting and thwarting any attempts, and, if unsuccessful, to recover rapidly any stolen materials). Protective measures for the private and government sectors were further upgraded in 1974–1975.

Second, governmental procedures were established at the same time for periodic high-level attention to the continuing reviews, by the AEC (now by the Nuclear Regulatory Commission and the Energy Research and Development Administration), of possible improvements in the physical security and materials accounting areas.

Third, the year 1973 also witnessed the emergence of a significant governmental consensus. This took the form of a directive to the effect that, although the methods and types of protective measures in the government and private sectors need not be identical, the degree of protection in both sectors should be essentially equivalent for similar types and quantities of nuclear materials.[1]

[1] The significance of this consensus and directive derives from the fact that physical security measures in the government sector (although in need of improvement) were generally much more thorough and demanding than those required for the private sector before the upgrading actions in 1973. The issue of adequately protecting nuclear weapons and nuclear materials within the government sector, particularly in light of potential terrorist activities, has also been the subject of scrutiny and concern in recent years. Protection requirements for the government sector have been upgraded since 1973, and are continuing to be improved.

Fourth, the federal budget for domestic safeguards research, development, and application began to grow, although the AEC's proposed level of additional expenditures was cut by the administration on budgetary and technical grounds.

Fifth, by late 1974, policy had been established that the government's examination of any country's request for substantial quantities of high-enriched uranium would specifically include, as a condition for supply, a determination that the requesting country had adequate protective measures for nuclear materials comparable to those in the United States. Any requests for plutonium would also trigger this consideration. In a similar vein, the 1974 proposal to supply nuclear reactors and fuel to Egypt and Israel was conditioned, among other things, on their application of adequate protective measures.

Finally, the desirability and logic of more clearly separating the responsibility for developing and promoting a grown nuclear industry from the responsibility for regulating it, for both safety and safeguards, were fulfilled in late 1974. Congress passed and President Ford signed the Energy Reorganization Act which abolished the AEC while establishing the Nuclear Regulatory Commission, as an independent regulatory agency, and the Energy Research and Development Administration, a major component of which concentrates on nuclear energy development and promotion.

Although the United States has come a great distance in the past few years, it is recognized within and outside government that further questions and suggested safeguards approaches warrant consideration.[2] Necessarily left unanswered in Washington's review of domestic safeguards and controls, but hardly going unnoticed in the process, was a major question: How could the United States encourage or ensure the application of at least equally effective protection measures for nuclear materials in other countries and in international transit? A few countries may provide more effective protection than the United States does, but physical security measures in many countries lag behind. Internationally accepted and monitored standards for protecting nuclear materials are nonexistent. Yet a diversion or theft in any country, or in transit between countries, could present a significant security risk not just to the country or countries immediately concerned but somewhere else as well.

Placing physical security conditions on United States exports of special nuclear materials and encouraging other nuclear exporters to apply similar conditions can get at major portions—but not nearly

[2] For examples of outstanding issues, see Willrich and Taylor, *Nuclear Theft*, pp. 169-172.

all—of the problem. Left unresolved are questions about nuclear materials not subject to any such export conditions, and questions of how to define "adequate" in the absence of international agreement on standards. Another approach, if widely accepted, could resolve much of the remaining problem on physical security. The United States announced in September 1974 that it would "urge the IAEA to draft an international convention for enhancing physical security" by setting forth "specific standards and techniques for protecting materials in use, storage, and transfer." [3]

The IAEA has come to take a more active interest in the physical protection of nuclear materials in recent years. A panel of experts, convened under IAEA auspices, produced the "grey book" in 1972. This work comprised recommendations on general and specific measures necessary to minimize the opportunities for any nuclear diversion or theft and to recover rapidly any stolen, or otherwise unaccounted for, nuclear material. However, in contrast to the agency's functions for safeguards against national nuclear diversion, the IAEA has "no responsibility" for providing, supervising, controlling, or implementing any physical protection system, and can give advice or assistance only at the request of a country.[4] The responsibility for protection against subnational threats resides, for all practical purposes, entirely with individual national authorities.

The requirements for basic materials accounting and reporting procedures, in conjunction with the IAEA safeguards system, also help guard against possible subnational diversion or theft. But they are designed with national diversion in mind and cannot hit the fundamental point of subnational scenarios. The key to providing assurances against subnational threats is a system of effective physical security measures including locks, barriers, alarms, guards, special transport vehicles or containers, communications networks, and contingency plans.

The issue is not that the IAEA should have more responsibility for ensuring or overseeing effective national physical protection systems, and certainly not that it (instead of national authorities) should have primary responsibility for such systems. Rather, the issues are: (1) physical security measures in many, if not most, other countries fall short of the guidelines and requirements suggested by the IAEA's panel of experts (as did those in the United States before

[3] Kissinger, "An Age of Interdependence," in *Department of State Bulletin*, vol. 71, no. 1842, p. 501.

[4] See International Atomic Energy Agency, *Recommendations for the Physical Protection of Nuclear Material* (Vienna: IAEA, June 1972), pp. 2-3.

measures were upgraded in 1973), not to mention their falling short of what the United States considers necessary to provide reasonable assurance against subnational threats; (2) some experts believe that measures in addition to those suggested by the IAEA and those established within the United States are yet necessary to provide adequate assurances; and (3) in an age of terrorist activity respecting no boundaries, the discrepancies between national physical security programs cannot be explained by objectively different implications for requirements in different societies. In this light, widely accepted international standards and methods for protecting nuclear materials could fill a substantial gap, and the IAEA with its expertise could play an important part.

However obvious the need may be to some, it cannot be assumed that broad international agreement and improvements will come easily. Significant improvements would cost only a very small percentage of the overall outlays for nuclear energy, but they would still involve cost. Other difficulties between a proposal for an international convention and its coming into effect include the concerns of some countries about what they consider encroachments on their sovereign prerogatives, divergent views about what measures are needed in different political-social systems, and different perspectives on the subnational threat.

Assumptions regarding the probability of damage or danger occurring frequently determine the priority given to preventive programs. The probability of significant subnational diversion, theft, or sabotage cannot be determined with any precision and may in some instances be quite low. But this does not diminish the potential severity of such a threat and the social, political, human, and other costs if it were carried to execution. The potential severity of the threat, not the probability of its occurring, argues most for making the odds as high as possible, worldwide, against subnational nuclear diversion. Stringent physical security and materials accounting measures should be viewed, much like international safeguards, not as an extra but as an inherent cost of nuclear energy.

Strengthening and Sustaining IAEA Safeguards

The development of IAEA safeguards for various steps in the nuclear fuel cycle has come about through always complicated and sometimes contentious negotiations spanning the years from the late 1950s. Detailed safeguards arrangements have been developed for nuclear reactors, fuel conversion and fabrication plants, and reprocessing facilities. Procedures for safeguarding uranium enrichment facilities are currently under consideration in the IAEA.

The objective of IAEA safeguards "is the timely detection of diversion of significant quantities of nuclear material from peaceful nuclear activities to the manufacture of nuclear weapons or of other nuclear explosive devices or for purposes unknown, and deterrence of such diversion by the risk of early detection."[5] The main features of the system comprise requirements for design review of facilities (to ensure that accounting measures and controls can be applied), for national record keeping or materials accounting, for reports to the IAEA based on these national records, and for periodic inspection visits by IAEA technical experts. The frequency of reports to and inspections by the IAEA will vary for different nuclear facilities and materials, depending upon their potential military significance.

Whether IAEA safeguards, based as they are primarily on national materials accounting records, can effectively meet the limited objective of detecting national diversion, and thereby deterring it, remains subject to question. Doubts will likely increase as the quantities of nuclear materials and the kinds or numbers of nuclear facilities in a country grow, especially since "significant quantities of nuclear materials" need not be much in some countries.

These doubts, however, do not necessarily lead to a conclusion that a major upgrading of the IAEA safeguards system itself is in order. Any significant upgrading, except in the important area of technical improvements, would be of dubious cost-effectiveness. For this and some much less warranted (if at all warranted) reasons, attempts to upgrade safeguards would run into considerable political opposition. Several developing countries believe that an increase in the IAEA's safeguards functions carries with it a price of a decrease in the promotion of nuclear energy for their economic development programs. Key industrial countries—such as West Germany, Italy, and Japan—have made it abundantly clear that the existing system is the most they consider acceptable given their concerns, among others, about protecting proprietary information and their competitive posture.

Regardless of questions about the effectiveness and limitations of IAEA safeguards, the system does establish the important principles of reporting to an international body and of inspection, including some right of "independent measurements and observations," by

---

[5] International Atomic Energy Agency, *The Structure and Content of Agreements between the Agency and States Required in Connection with the Treaty on the Non-Proliferation of Nuclear Weapons*, INFCIRC/153 (Vienna: IAEA, May 1971), p. 9.

international technical experts.[6] In today's context, acceptance of safeguards by countries can signal or reinforce a declaration of intention not to acquire nuclear explosive capabilities. This signal could, in turn, reduce the possibility of destabilizing suspicions among its neighbors or other countries. Acceptance and implemention of the precedent-setting principles in international safeguards could also provide a basis for further IAEA responsibilities and functions if sufficient international support should develop and demand these in the future.

In sum, the application of IAEA safeguards and the assurances they provide, even though obviously limited, are far preferable to no international safeguards at all, and the value of extending their application should not be discounted. Would not Egypt, for example, at least have less cause for suspicion were the Israeli reactor at Dimona subject to international safeguards? Or what about Pakistan if India were to accept appropriate international observations for its nuclear explosive program and safeguards on all its other nuclear activities? Although significant improvements can come through technical advances such as better measuring techniques, the primary path for strengthening international safeguards lies in extending their application on the broadest possible basis. The obligations set forth in the Nuclear Nonproliferation Treaty already cover a great portion of the international spectrum. But issues remain, as previously noted, that are especially relevant to nonparties to that treaty.

Sustaining the current safeguards system will be as important, and could be as difficult, as the task of gaining broader acceptance. Costs for safeguarding nuclear activities in those countries obligated or offering to accept their application will rise sharply in the years ahead to keep pace with nuclear energy growth. Estimated costs reach $50 million by 1985, with substantial increases continuing thereafter. Manpower, technical training, and equipment are key areas where increases will be required to sustain standard safeguards operations. The present is none too soon, therefore, to give serious consideration to how the necessary funding can best be ensured and equitably allocated.[7]

---

[6] Ibid., p. 3. It is worth noting that, in the event significant quantities of nuclear material subject to safeguards cannot be accounted for in a country, the IAEA statute provides for certain procedures and rights. In cases of unremedied noncompliance, the IAEA will inform its members, the United Nations General Assembly, and the Security Council; further, it can suspend IAEA assistance, call for the return of nuclear materials and equipment supplied to the country in question, and suspend it from IAEA membership.

[7] It has been suggested that a "safeguards insurance premium," whereby each user nation would contribute in proportion to its nuclear energy capacity, might

Today, with considerable attention in the United States and a few other countries focused on problems of proliferation, there is reasonable room for hope regarding additional political and financial support for the IAEA generally and specifically for its safeguards operations. But the room for hope is not very wide, and even it could change in the years ahead with a rapidly growing nuclear power industry. Even though safeguards costs are conspicuously modest in comparison to overall nuclear energy costs, the "spread of civilian nuclear power in a commercially competitive global environment threatens to outrun the willingness of national governments concerned to accept safeguards and to open peaceful nuclear activities . . . to international inspection."[8] Without increased political and financial support, a gradual deterioration in the quality and quantity of safeguards could occur in times which call for just the opposite.

A Word about Peaceful Nuclear Explosions

The United States has fostered not only the idea of commercially competitive generation of electricity through nuclear power, but also the potential benefits of peaceful nuclear explosions (PNEs). First came the idea of nuclear excavation for the construction of harbors and canals, and then that of underground detonations for the recovery of gas and oil. For years, "atoms for peace" included awaiting the benefits of PNEs.

While the United States and many other countries have generally come to accept the benefits of generating electricity by nuclear energy as outweighing the problems posed by it, the same is not true of PNEs. Washington's hopes for their commercial application have faded significantly and rapidly since the late 1960s. A nadir was reached in 1974 when Congress acted to prohibit the use of energy research and development funds for any field testing of nonmilitary nuclear devices. Today, the promise of PNEs is, for the most part, considered a "dud" in the United States.

The picture on the international scene is much more mixed. Among the other nuclear-weapon states, the Soviet Union conducts

meet this objective. Regardless how one judges the merits of this suggestion, it is becoming ever clearer that (1) the present system of dividing costs among IAEA members according to the standard United Nations formula will not likely meet this objective without substantial voluntary contributions; and (2) these voluntary contributions may not provide the best assurances for sustaining adequate financial support. See Doub and Dukert, "Making Nuclear Energy Safe and Secure," *Foreign Affairs*, vol. 53, pp. 763-764.

[8] Willrich and Taylor, *Nuclear Theft*, p. 190.

the most active and varied PNE program in the world. Besides specific excavation and underground applications, Moscow has plans for a large-scale excavation project, involving a multitude of shots, to join the Kama and Pechora rivers with the aim of replenishing a receding Caspian Sea. France has indicated an interest in some PNE applications and in becoming a supplier of PNE services to other countries. Great Britain, on the other hand, has no plans for an active PNE program or for providing any services abroad; and the People's Republic of China is not known to have any interest in PNE applications. Only a handful of non-nuclear-weapon countries has contacted the IAEA, U.S. governmental agencies, or U.S. and other private firms for technical studies of the feasibility and desirability of specific PNE applications.[9] However, one country (India) has demonstrated, and two others (Brazil and Argentina) have expressed, an interest in developing nuclear explosives for some (as yet largely undefined) peaceful purposes.

The United States frequently finds itself pulled in at least two directions within this picture. It has only a trickle of its previous PNE interest and program, along with a concern that any shots (generally excavation shots) which resulted in radioactive debris exiting the borders of the detonating country would be in contravention of the 1963 Limited Test Ban Treaty. This diminished interest and this concern are translated into no enthusiasm for encouraging PNE applications, for providing any PNE services to others, or for opening the Pandora's Box of such issues as reinterpretation or amendment of the Limited Test Ban Treaty.

On the other hand, Article V of the NPT obligates the United States to make available, under international observation and procedures, "potential benefits from any peaceful applications of nuclear explosions." Washington has supported and initiated some efforts within the IAEA pertaining to the establishment of international guidelines and procedures for any future PNE services. Besides decisions dealing with specific operational matters, such as whether to participate in feasibility studies for possible PNE applications abroad, further consideration of international arrangements is about

[9] The most talked about are the Egyptian inquiry through West Germany of the possibility of using PNEs to excavate a canal from the Mediterranean to the Qattara Depression for generating hydroelectric power and generally developing the region; the examination by U.S. firms of the advantages and disadvantages of using PNEs to help in the construction of a canal across Thailand's Kra Isthmus in order to shorten the oil route to East Asia; and, earlier, Australia's consideration of the application of PNEs for harbor construction at Cape Kerauden.

the most to be expected of general United States policy in this area, at least for the near term.

Two basic queries regarding the implications of this limited policy scope for nonproliferation interests deserve examination. First, does Washington's lack of enthusiasm and initiatives for realizing international PNE services actually encourage or in any way facilitate the development of nuclear explosive capabilities abroad? Second, given Washington's relative disinterest—indeed, definite lack of enthusiasm—would it not be advisable to put further emphasis on discouraging the whole idea of PNEs?

Putting the first question in more specific and frequently phrased terms, do Washington's position and the absence of international PNE services make a difference relative to present or potential impulses in other countries toward PNE development, either (1) by lending credibility to the arguments (whether used to justify a position internationally or within a country's political and bureaucratic processes) which maintain a need to keep open or exercise an indigenous PNE option, or (2) by failing to provide an argument (international PNE services could be cheaper and easier) for possible forces against indigenous development within a country?

It goes without saying that the availability of international PNE services might influence the way the issue of PNE development would be debated within a country. But given potentially available sources of supply, and the possibility of international services if the matter were really pressed, a non-nuclear-weapon country interested strictly in a specific PNE project or projects would surely not have to be stimulated toward indigenous development by the absence of ready international arrangements. At any rate, no such country has pushed the point. On the other hand, a country interested in PNEs for a mixture of perceived political, prestige, technological, and economic reasons would not likely find international services a very attractive alternative to national programs, and certainly would not if implicit security-military ingredients were blended into its motivation.

These broad assumptions make it hard to see how the absence of international services and Washington's lack of enthusiasm for them and for PNEs generally would encourage or facilitate PNE developments abroad. The absence of international services may lend credibility to the argument that this lack necessitates keeping open or exercising a national PNE option, but the credibility will remain quite limited. Moreover, the alternative of significantly increased support for international PNE services could present problems of its own, particularly if it were combined with active PNE development in the United States.[10]

The second basic question naturally follows from Washington's relative disinterest in PNEs: Would it not be advisable to put further emphasis on discouraging the whole idea of peaceful applications of nuclear explosives? In some respects, a step has been taken in this direction. The "U.S. Statement on Nuclear Issues" in October 1974 contained a degree of candor not often found in previous official pronouncements on the subject. It noted that "the commercial utility of PNEs has not been proved" to date, that their use "is a highly complicated matter politically and legally," and that the international review of them should include "their potential limitations as well as their potential benefits." [11]

The legitimacy of other countries' interests in possible PNE applications—however poorly founded in Washington's eyes or in fact—cannot simply be denied. (After all, it took Washington several years, many studies, and a few tests before it fell off its hopes.) Questions now before the IAEA and any requests for PNE services should receive balanced analysis, and U.S. policy and participation should be geared toward that end. But more active discouragement would not likely turn off existing interests in PNEs and, especially if it also meant less subtle discouragement, could prove an irritant in relations with some countries.

Despite the points noted in the foregoing paragraphs, the desirability of a more positive approach by the United States toward international arrangements for PNE services cannot be foreclosed, if only to ensure adequate international guidelines and procedures. The political-legal logic of events, if not any demonstrated benefits of PNEs, may continue to move in the direction of such arrangements. Precisely because of the points noted in the foregoing paragraphs, however, more active policy options need to be judged primarily on their own merits, and not in terms of any expected nonproliferation spin-offs.

---

[10] There is a school of thought which maintains that the United States would require an active domestic program for the commercial development of PNEs in order to support international services. Such a program would at least imply expectations of substantial benefits. These expectations would, of course, be contrary to Washington's current assessment of the value of PNEs for the United States, if not generally. But if this assessment and the concomitant policy were to change toward strong support, other countries' interests in PNEs might well increase. Even if international services were made available on reasonable terms, independent programs and capabilities may appear more attractive if the pay-offs are seen to be high.

[11] Symington, "U.S. Statement on Nuclear Issues," p. 7.

Difficult PNE questions will continue to confront Washington not only in the areas described above but also in nuclear test ban issues and negotiations. Differences over PNEs complicated the negotiation of the Threshold Test Ban Treaty between Washington and Moscow in the summer of 1974, and were yet to be resolved after over a year of further discussion.[12] Indeed, given the two capitals' varying interests on the subject and their views on verification, PNE questions could not easily be settled in this or any other test ban negotiation. In more ways than one, PNEs promise to remain a thorny issue.

[12] The United States preferred to include all PNEs under the 150-kiloton limitation of the Threshold Test Ban. With its plans for the Kama-Pechora canal project, envisaging several excavation shots above 150 kilotons and altogether a multitude of detonations, "it was at Soviet insistence that the . . . Treaty left open the question of peaceful explosives for subsequent negotiations." See Fred C. Iklé, "The Dilemma of Controlling the Spread of Nuclear Weapons While Promoting Peaceful Technology," address before Duke University Law Forum (18 September 1974), in *Department of State Bulletin*, vol. 71, no. 1843 (21 October 1974), p. 545.

By mid-1975, some progress (but no definite conclusions or agreement) had reportedly been made in these negotiations, including progress in the area of on-site inspection procedures (which might set an important precedent in the overall context of United States-Soviet Union arms control).

CONCLUSION

Until recent years, international efforts directed toward the problem of safeguarding nuclear materials were focused primarily on covert diversion by national authorities for military purposes. This aspect deserves continuing attention but not a dominant position among the security risks of the peaceful atom. It has been apparent for some time that equally (and probably more) substantial parts of the problem include: (1) acquisition through peaceful programs of the technical and material preconditions for developing nuclear weapons, possibly in a relatively short time once these preconditions are obtained; (2) overt diversion or takeover of nuclear materials and facilities by national authorities for military purposes; (3) development of a nuclear explosive capability labeled "for peaceful purposes"; and (4) diversion or theft of nuclear materials, or sabotage of nuclear facilities, by dissident individuals or terrorist groups.

The approaches, controls, safeguards, and protective measures suggested in this essay take into account these various aspects of the dilemma of the peaceful atom. Taken together, the proposals can still only reduce the proliferation and other security risks of expanding nuclear energy industries. If all the proposed risk-reducing measures and more were fully implemented by nuclear suppliers and users within the international community, some risks of national or subnational diversion would remain wherever there are nuclear materials and facilities. A country determined to possess the preconditions for a nuclear explosive capability or to possess actual nuclear weapons could acquire either with time and resources. A determined individual or group could succeed in capturing nuclear materials by stealth or by force. Airtight or absolute assurances might appear feasible in theory, but they would not be compatible with any nuclear energy industry.

The impossibility of eliminating all potential dangers does not devalue the importance of lessening, as much as possible, the chances for their occurrence. Nor does the persistence of some risk diminish the value of attempting to influence the pace at which the world might have to deal with new nuclear powers. The proposed reduction-of-risk measures are worth pursuing even if they, added to the existing structure of restraints, only postpone the day when any given country can or does go nuclear.

The buying of time, or delay, represents a not insignificant policy product for several reasons. The nuclear option could appear less attractive and more risky the longer the time required between decision and weapon production. Time, resource expenditure, and possible difficulties could be on the side of internal forces working against a nuclear impulse. Time itself can work changes in internal and external conditions, power structures, perspectives, and policies. Last, but not least, only a short lead-time capability provides a nuclear option in conflict or crisis situations.

In the final analysis, nonproliferation efforts and objectives will be served, or disserved, by much more than the specific suggestions considered in this essay for reducing the risks directly associated with the worldwide diffusion of nuclear energy programs. They will be served, and far more consequentially in most respects: (1) by a pattern of foreign policy programs, a defense posture, a network of security relationships and guarantees, and general strategies designed to develop, maintain, or enhance conditions for regional and global stability (conditions wherein the disincentives for nuclear acquisition will be reinforced, and other countries will feel less need to move in that direction); and (2) by specific policies, strategies, and tactics toward individual countries, which by design or otherwise help to reduce any possible incentives and reinforce any disincentives to their acquiring nuclear explosive capabilities.

Nonproliferation objectives can also be served by measures such as regional nuclear-free zones,[1] similar to that established under the Latin American Nuclear Free Zone Treaty, and indirectly by other types of nuclear arms control agreements. For the most part, such developments will contribute to nonproliferation aims only so long as they neither bring into question the capability and will behind

[1] Pursuant to a United Nations resolution in 1974, a group of governmental experts recently concluded a study of this subject under the auspices of the Conference of the Committee on Disarmament. See *Comprehensive Study of the Question of Nuclear-Weapon-Free Zones in All of Its Aspects*, CCD/467 (18 August 1975).

security guarantees, nor contain significant PNE "loopholes" for proliferation.

The two basic categories of policy given above are especially critical. Generally speaking, a country's decision to acquire nuclear devices will be driven not by the relative ease or difficulty accompanying the task, but rather by the intent of its leaders based on their perceptions of national self-interest. The furthering of conditions conducive to political restraint on the nuclear question will become increasingly critical as the spread of nuclear materials, equipment, and technology, under peaceful programs, continues to erode technical and economic barriers to nuclear proliferation. The proliferation problem has always been, and will remain, only one part of a larger political-security puzzle.

Many facets, indeed, surround the problem of nuclear proliferation. How to handle it becomes a much more complicated process in an age of peaceful nuclear plenty than it was in an age of nuclear scarcity. Controlling the dissemination of the preconditions for "going nuclear," and safeguarding whatever spread occurs against national and subnational diversion, are two central requirements for reducing the security risks of the peaceful atom. If the overall picture of multinational approaches, technology controls, safeguards, and other protective measures resembles an unglamorous assemblage of less than "half-loaves," a challenge remains to devise enough so that, taken together in the broader context, they will form a sufficient structure of restraint and assurances.

APPENDIX

Parties to the Nuclear Nonproliferation Treaty

Ratifications Deposited

Afghanistan (1970)
Australia (1973)
Austria (1969)
Belgium (1975)
Bolivia (1970)
Botswana (1969)
Bulgaria (1969)
Cameroon (1969)
Canada (1969)
Chad (1971)
China, Republic of (1970)
Costa Rica (1970)
Cyprus (1970)
Czechoslovakia (1969)
Dahomey (1972)
Denmark (1969)
Dominican Republic (1971)
Ecuador (1969)
El Salvador (1972)
Ethiopia (1970)
Finland (1969)
Gambia (1975)
German Democratic Republic (1969)
Germany, Federal Republic of (1975)
Ghana (1970)
Greece (1970)
Guatemala (1970)
Haiti (1970)
Honduras (1973)
Hungary (1969)
Iceland (1969)
Iran (1970)
Iraq (1969)
Ireland (1968)
Italy (1975)
Ivory Coast (1973)
Jamaica (1970)
Jordan (1970)
Kenya (1970)
Korea, Republic of (1975)
Laos (1970)
Lebanon (1970)
Lesotho (1970)
Libya (1975)
Liberia (1970)
Luxembourg (1975)
Madagascar (1970)
Malaysia (1970)
Maldives, Republic of (1970)
Malta (1970)

Mali (1970)
Mauritius (1969)
Mexico (1969)
Mongolia (1969)
Morocco (1970)
Nepal (1970)
Netherlands (1975)
New Zealand (1969)
Nicaragua (1973)
Nigeria (1968)
Norway (1969)
Paraguay (1970)
Peru (1970)
Philippines (1972)
Poland (1969)
Romania (1970)
San Marino (1970)
Senegal (1970)
Somalia (1970)
Sudan (1973)
Swaziland (1969)
Sweden (1970)
Syrian Arab Republic (1969)
Togo (1970)
Tunisia (1970)
U.S.S.R. (1970)
United Kingdom (1968)
United States (1970)
Upper Volta (1970)
Uruguay (1970)
Vietnam (South) (1971)
Yugoslavia (1970)
Zaire (1970)

Accessions Deposited

Burundi (1971)
Central African Republic (1970)
Gabon (1974)
Holy See (1971)
Khmer Republic (1972)
Sierra Leone (1975)
Thailand (1972)
Western Samoa (1975)
Rwanda (1975)

Notification of Succession

Fiji (1972)
Tonga (1971)

SELECTED BIBLIOGRAPHY

Public Documents

Conference of the Committee on Disarmament. *Comprehensive Study of the Question of Nuclear-Weapon-Free Zones in All of Its Aspects.* Report of the Ad Hoc Group of Governmental Experts. Geneva: CCD/467, 18 August 1975.

International Atomic Energy Agency. *Recommendations for the Physical Protection of Nuclear Materials.* Panel of Experts Report to the IAEA. Vienna: IAEA, June 1972.

———. *The Structure and Content of Agreements between the Agency and States Required in Connection with the Treaty on the Non-Proliferation of Nuclear Weapons.* INFCIRC/153. Vienna: IAEA, May 1971.

U.S. Arms Control and Disarmament Agency. *International Negotiations on the Treaty for the Nonproliferation of Nuclear Weapons.* ACDA Publication No. 48, January 1969.

U.S. Atomic Energy Commission. *Draft, Generic Environmental Statement Mixed Oxide Fuel: Recycle Plutonium in Light Water-Cooled Reactors.* Vol. 1 (Summary and Conclusions). AEC/WASH-1327, August 1974.

———. *Nuclear Power Growth: 1974–2000.* AEC/WASH-1139, February 1974.

U.S. Congress, Joint Committee on Atomic Energy. *Hearings on S. 3323 and H.R. 8862 to Amend the Atomic Energy Act of 1946.* 83rd Cong., 2d sess., 1954.

U.S. Department of State. *Bulletin.* Vols. 14, 29, 30, 71, 72, and 73.

———. *Documents on Disarmament 1945–1959.* Vol. 1. Washington, D. C.: United States Government Printing Office, 1960.

U.S. Senate, Committee on Foreign Relations. *Hearings on the Nonproliferation Treaty.* 90th Cong., 2d sess., 1968.

Books

Bader, William B. *United States and the Spread of Nuclear Weapons.* New York: Pegasus, 1968.

Barnaby, C. F., ed. *Preventing the Spread of Nuclear Weapons.* Pugwash Monograph 1. New York: Souvenir, 1969.

Beaton, Leonard. *Must the Bomb Spread?* Harmondsworth, Middlesex: Penguin, 1966.

Beaton, Leonard, and Maddox, John. *The Spread of Nuclear Weapons.* New York: Praeger, for the Institute of Strategic Studies, 1962.

Bechhoefer, Bernhard G. *Postwar Negotiations for Arms Control.* Washington, D. C.: The Brookings Institution, 1961.

Boskey, Bennett, and Willrich, Mason, eds. *Nuclear Proliferation: Prospects for Control.* New York: Dunellen, 1970.

Brzezinski, Zbigniew. *The Fragile Blossom: Crisis and Change in Japan.* New York: Harper Torchbooks, 1972.

Buchan, Alastair. *A World of Nuclear Powers?* Englewood Cliffs, N.J.: Prentice-Hall, 1966.

Dougherty, James E. *How to Think about Arms Control and Disarmament.* New York: Crane, Russak, and Company, for the National Strategy Information Center, 1973.

Fischer, Georges. *The Non-Proliferation of Nuclear Weapons.* Translated by David Wiley. London: Europa Publications, 1971.

Kramish, Arnold. *The Peaceful Atom and Foreign Policy.* New York: Harper and Row, 1963.

Nau, Henry R. *National Politics and International Technology: Nuclear Reactor Development in Western Europe.* Baltimore: Johns Hopkins University Press, 1974.

Quester, George H. *The Politics of Nuclear Proliferation.* Baltimore: Johns Hopkins University Press, 1973.

Roberts, Chalmers M. *The Nuclear Years: The Arms Race and Arms Control 1945–70.* New York: Praeger, 1970.

Rosencrance, Richard N., ed. *The Dispersion of Nuclear Weapons: Strategy and Politics.* New York: Columbia University Press, 1964.

Stockholm International Peace Research Institute. *Nuclear Proliferation Problems.* Stockholm: Almqvist and Wiksell, 1974.

Williams, Shelton L. *The U.S., India, and the Bomb.* Baltimore: Johns Hopkins University Press, 1969.

Willrich, Mason, ed. *Civil Nuclear Power and International Security.* New York: Praeger, 1971.

Willrich, Mason. *Global Politics of Nuclear Energy.* New York: Praeger, 1971.

Willrich, Mason, and Taylor, Theodore B. *Nuclear Theft: Risks and Safeguards.* A Report to the Energy Policy Project of the Ford Foundation. Cambridge, Mass.: Ballinger, 1974.

Articles, Reports, Statements, and Studies

Acheson, Dean, and Lilienthal, David E. "A Report on the International Control of Atomic Energy (17 March 1946)." *Department of State Bulletin,* vol. 14, no. 353 (7 April 1946), pp. 553–560.

Baruch, Bernard. "The Baruch Plan." Statement before the United Nations Atomic Energy Commission (14 June 1946). *Department of State Bulletin,* vol. 14, no. 364 (23 June 1946), pp. 1057–1062.

Bloomfield, Lincoln P. "Nuclear Spread and World Order." *Foreign Affairs,* vol. 53 (July 1975), pp. 743–755.

Doub, William O., and Dukert, Joseph M. "Making Nuclear Energy Safe and Secure." *Foreign Affairs,* vol. 53 (July 1975), pp. 756–772.

Dulles, John Foster. "Amending the Atomic Energy Act." Statement before the Joint Committee on Atomic Energy (3 June 1954). *Department of State Bulletin,* vol. 30, no. 781 (14 June 1954), pp. 926–928.

Eisenhower, Dwight D. "Atomic Power for Peace." Address before the United Nations General Assembly (8 December 1953). *Department of State Bulletin,* vol. 29, no. 756 (21 December 1953), pp. 847–851.

Gilinsky, V., and Hoehn, W. *On the Military Significance of Small Uranium Enrichment Facilities Fed with Slightly Enriched Uranium.* RAND, RM-6123-ARPA, January 1970.

———. *Nonproliferation Treaty Safeguards and the Spread of Nuclear Technology.* RAND, R-501, May 1970.

Halperin, Morton H. "A Ban on the Proliferation of Nuclear Weapons." In *First Steps to Disarmament*, edited by Evan Luard, pp. 132–160. New York: Basic Books, 1965.

Iklé, Fred C. "The Dilemma of Controlling the Spread of Nuclear Weapons while Promoting Peaceful Technology." Address before Duke University Law Forum, Durham (18 September 1974). *Department of State Bulletin*, vol. 71, no. 1843 (21 October 1974), pp. 543–547.

———. "Nth Countries and Disarmament." *Bulletin of the Atomic Scientists*, vol. 16 (December 1960), pp. 391–394.

Iklé, Fred C.; Pollock, Herman; and Sober, Sidney. "Proposals to Export Nuclear Technology to Egypt and Israel." Statements before the Subcommittees on International Organizations and Movements and on the Near East and South Asia of the House Committee on Foreign Affairs (9 July 1974). *Department of State Bulletin*, vol. 71, no. 1832 (5 August 1974), pp. 248–254.

Joint Communiqué, issued at the conclusion of Secretary of State Kissinger's visit to New Delhi, India. Department of State Press Release, no. 449 (30 October 1974).

Kissinger, Henry A. "An Age of Interdependence: Common Disaster or Community." Address before the United Nations General Assembly (23 September 1974). *Department of State Bulletin*, vol. 71, no. 1842 (14 October 1974), pp. 498–504.

———. "Building International Order." Address before the United Nations General Assembly (22 September 1975). *Department of State Bulletin*, vol. 73, no. 1894 (13 October 1975), pp. 545–553.

———. "Toward a Global Community: The Common Cause of India and America." Address before the Indian Council on World Affairs, New Delhi (28 October 1974). *Department of State Bulletin*, vol. 71, no. 1848 (25 November 1974), pp. 740–746.

Kramish, Arnold. "The Watched and the Unwatched: Inspection in the Nonproliferation Treaty." *Adelphi Papers*, no. 36 (June 1967), Institute for Strategic Studies, London.

Pranger, Robert J., and Tahtinen, Dale R. *Nuclear Threat in the Middle East*. Foreign Affairs Study No. 23. Washington, D. C.: American Enterprise Institute for Public Policy Research, 1975.

Quester, George H. "Can Proliferation Now Be Stopped?" *Foreign Affairs*, vol. 53 (October 1974), pp. 77–97.

Scheinman, Lawrence. "Security and a Transnational System: The Case of Nuclear Energy." In *Transnational Relations and World Politics,* edited by Robert O. Keohane and Joseph S. Nye, Jr., pp. 276–299. Cambridge, Mass.: Harvard University Press, 1972.

Sisco, Joseph J. "Department Discusses Proposed Nuclear Reactor Agreements with Egypt and Israel." Statement before the Subcommittees on International Organizations and Movements and on the Near East and South Asia of the House Committee on Foreign Affairs (16 September 1974). *Department of State Bulletin,* vol. 71, no. 1841 (7 October 1974), pp. 484–486.

Symington, Stuart. "U.S. Statement on Nuclear Issues." Address before the First Committee of the United Nations General Assembly (21 October 1974). Press Release by Senator Symington's Office.

"Text of Communiqué," issued following the meetings of President Ford and President Giscard d'Estaing in Martinique (16 December 1974). *Department of State Bulletin,* vol. 72, no. 1855 (13 January 1975), pp. 42–43.

Young, Elizabeth. "The Control of Proliferation: The 1968 Treaty in Hindsight and Forecast." *Adelphi Papers,* no. 56 (April 1969), Institute for Strategic Studies, London.

Cover and book design: Pat Taylor